DESIGNING GREEN CEMENT PLANTS

DESIGNING GREEN CEMENT PLANTS

S. P. DEOLALKAR

Author of
Handbook for Designing Cement Plants
&
Nomograms for Design & Operation of Cement Plants

ELSEVIER

AMSTERDAM • BOSTON • HEIDELBERG • LONDON
NEW YORK • OXFORD • PARIS • SAN DIEGO
SAN FRANCISCO • SINGAPORE • SYDNEY • TOKYO
Butterworth-Heinemann is an imprint of Elsevier

BSP **BS Publications**
A Unit of BSP Books Pvt. Ltd., India.

Butterworth Heinemann is an imprint of Elsevier
The Boulevard, Langford Lane, Kidlington, Oxford OX5 1GB, UK
225 Wyman Street, Waltham, MA 02451, USA

Notices

Knowledge and best practice in this field are constantly changing. As new research and experience broaden our understanding, changes in research methods, professional practices, or medical treatment may become necessary.

Practitioners and researchers must always rely on their own experience and knowledge in evaluating and using any information, methods, compounds, or experiments described herein. In using such information or methods they should be mindful of their own safety and the safety of others, including parties for whom they have a professional responsibility.

To the fullest extent of the law, neither the Publisher nor the authors, contributors, or editors, assume any liability for any injury and/or damage to persons or property as a matter of products liability, negligence or otherwise, or from any use or operation of any methods, products, instructions, or ideas contained in the material herein.

Library of Congress Cataloging-in-Publication Data
A catalog record for this book is available from the Library of Congress

British Library Cataloguing in Publication Data
A catalogue record for this book is available from the British Library

For information on all Butterworth Heinemann publications visit our website at http://store.elsevier.com/

ISBN: 978-0-12-803420-0

DEDICATION

This, my third book,
is for
Urmila,
my lifelong partner
and
proprietor of
Deolalkar Consultants

CONTENTS

Supplementary materials for this book will be available in
http://www.booksite.elsevier.com/9780128034200/

FOREWORD

Cement use is a measure of a country's progress; it has a direct correlation with GDP. However, the cement manufacturing process itself has been a cause for concern because it creates severe pollution as measured by CO_2 emissions. Globally, the cement industry accounts for 5% of the world's CO_2 emissions. The basic cause is of course the use of limestone as a raw material and the fossil fuel coal as a fuel, which together contribute about 0.8 tons of CO_2 emissions per ton of clinker.

As the search for alternate raw materials and fuels is still in its infancy, particularly in India, and the need for economic growth is an essential prerequisite for a growing population, the dependence on natural resources is so rigorous that the cement industry's contribution to sustainable development appears to be a far distant dream.

Therefore, the cement industry needs to urgently look for alternatives not only because of economic but also because of ecological aspects. While cement manufacturers are continuously striving to achieve more efficient and environmentally friendly production methods, it will require innovative efforts at every point in the process, starting from designing a cement plant to installing it and operating it, to control overall CO_2 emissions. It is in this respect that Mr. S.P. Deolalkar's book *Designing Green Cement Plants* will serve the industry immensely as a ready resource by which to select and install the correct process of cement manufacture.

I have known Mr. Deolalkar for many years, and his perseverance and assiduity in making very valuable contributions to the cement industry are greatly appreciated. In fact his first book, *Handbook for Designing Cement Plants*, continues to be an Gita (essential text) for cement technologists.

I am sure that this new book, *Designing Green Cement Plants*, will also soon be seen on the desks of cement entrepreneurs, planners, and project managers as a ready guide for the selection of green cement manufacturing processes and practices.

I have had the privilege of association with the Confederation of Indian Industry (CII) in spearheading a movement to "green" the cement industry for almost a decade now, and it is very gratifying to see Mr. Deolalkar giving this movement a push by putting down his ideas and experience in designing green cement plants.

I look forward to his book's success in proving that "greening" measures in the cement industry will prove to be sustainable as well as good business. My best wishes to him.

G. Jayaraman
Chairman, Cementech-CII
10 May 2013
Kolkata

PREFACE

My first book, *Handbook for Designing Cement Plants,* was published in 2008. However, I had begun working on it in 2002. Thus, more than a decade has passed since that book was written. During this period, momentous changes have taken place in the cement industry in terms of the growth and content of cement plants.

India has jumped to the second position in the world in terms of installed capacity and production of cement. The capacity of a single line has increased from 3000 tons per day (tpd) of clinker at the end of the twentieth century to more than 10,000 tpd of clinker in the first decade of the twenty-first century.

During this time, the world's attention was pointedly drawn to the "warming effect" of "greenhouse" gases (GHGs), of which carbon dioxide (CO_2) is the principal constituent. The cement industry, which contributes substantially to the generation of CO_2 through the cement-making process, therefore was compelled to take note of this effect. It began taking steps to reduce GHG emissions.

Emphasis had also shifted during this period from growth to "sustainable development."

While cement plants were doing everything possible to improve their operational efficiencies (mainly because of economic considerations), a new dimension was thus added: to improve operation of cement plants in terms of environmental friendliness and sustainability in addition to reducing dust and noise pollution.

The process of manufacturing cement from clinker involves the release substantial quantities of CO_2, the principal GHG. From the ~1.55 kg of raw meal required to make 1 kg of clinker, the CO_2 released is ~0.51 kg. Burning a fossil fuel (such as coal) during the clinkering process releases an additional ~0.29 kg of CO_2 per kg of clinker, making the total CO_2 emission ~0.80 kg/kg of clinker or 0.77 kg/kg of OPC (ordinary portland cement).

It is estimated that the cement industry contributes 5% of the total GHG emissions.

In 1990, the nations of the world came together to set targets to reduce GHG emissions. India did not lag behind but set a target to reduce GHG emissions from the present level of $\sim$700 kg/t cement to $\sim$560 kg in the next 20 years, a reduction of 20%.[1]

It is thus imperative to examine the various possibilities for reducing GHG emissions from new and existing cement plants to levels which are within the set targets.

The relatively new concern about **Sustainability** touches on the conservation of natural resources used in making cement. It is necessary, therefore, to find ways to reduce the use of limestone and fossil fuels in the cement-making process.

Cement plants that are designed taking these important aspects into account are aptly called "green" cement plants.

Therefore, new cement plants must have features that were perhaps optional for older plants, such as:
- maximum production of blended cements
- facilities for using alternate fuels
- facilities for waste heat recovery

Because they contribute to reducing GHG emissions and also to sustainability.

These plants should also consider making composite cements such as limestone-based/low-grade cements to conserve limestone reserves.

A recent development is the research on "cement substitutes." Several such products are currently in the pilot plant stage and show the promise of replacing (to some extent) cements as we know them today. These new products are in essence "green." Therefore, the cement industry should be extremely interested in their development.

In the past decade, there have also been several developments in the processes and machinery used to make cement.

There have been, in particular, developments in the preparation of alternative fuels, including biomass fuels, to make them suitable for firing in cement kilns. The consequent emphasis on reducing a number of undesirable elements from the atmosphere led to developments in kiln bypass systems. In addition, calciners were designed to reduce NO_X emissions.

Efforts are simultaneously being made to reduce GHG emissions by other routes.

---

[1] Low Carbon Technology Roadmap for the Indian Cement Industry. A publication of the Confederation of Indian Industries (CII), Green Business Centre, Hyderabad.

One route is to capture emitted CO_2 and store it safely (carbon capture and storage). Another route makes use of captured CO_2 to make new cement.

All these developments need to be taken into account when planning and designing new cement plants.

Also during the past decade, because of the increase in the sizes of cement plants and because of the passage of time, significant changes have taken place in the structures of capital costs and costs of production of cement.

Taking all these aspects into account, I felt that it was necessary to put the new developments aimed at making cement plants "green" and sustainable together in the form of a book for the convenience of cement plant designers and operators.

Hence, I composed this book, *Designing Green Cement Plants*. In a way, it is a sequel to my first book, *Handbook for Designing Cement Plants*.

The CII—Green Business Centre at Hyderabad, India, has spearheaded the movement for green cement and has helped me by allowing me to use their valuable publications, which promote the green movement. I am very grateful to them.

I am also grateful to all those connected with the cement industry (manufacturers, consultants, machinery designers, and machinery manufacturers) who helped me as I wrote this book. Their specific contributions have been acknowledged separately.

I hope that this new book receives the same warm response from the world as *Handbook for Designing Cement Plants* did.

S.P. Deolalkar

ACKNOWLEDGMENTS

I have received generous help from various quarters as I was writing this book. I am hereby making a sincere and humble attempt to acknowledge this help. It is, however, possible that there could be a few inadvertent omissions. I apologize for them.

I received the inspiration to write this book from the annual seminars on green cement organized by IGBC at Hyderabad. CII, the parent body of IGBC, has also brought out several publications on different aspects of green cement. I am grateful to CII for their permission to use material from their publications in this book.

Mr. G. Jayaraman, Chairman, Green Cemtech, CII—Godrej Green Business Centre, has been spearheading the movement for green cement. It is appropriate that he give this book his blessing by writing the Foreword. I thank him for that, and I thank all the executives of CII for their support.

Any book on cement plants cannot be authentic without the support of

1. Designers and manufacturers of cement machinery
2. Vendors of major auxiliaries
3. Cement companies
4. Cement consultants
5. Cement technologists

I approached leaders in each of these categories and sought their help by way of data, drawings and photos of machinery, plant layouts, and so on to include in this book. I sincerely thank all who have allowed me to do so. I have mentioned these sources under each figure or table and at the end of the book in Section Sources.

Principal among them are

FLSmidth Pvt Ltd

KHD Humboldt Wedag

Polysius AG Krupp Thyssen

Claudius Peters AG

IKN Engineering India (P) Ltd

Associated Cement Companies Ltd

ABG Cement Company Ltd

Bhavya Cement Company Ltd

Green cement plants must have systems for using alternate fuels (AF) and raw materials as well as systems for waste heat recovery. I have received help from concerned companies like GEPIL, who are doing a great job in making AF available to the cement industry and other industries and to makers of waste heat recovery systems such as Thermax and TESPL.

The Internet has been an excellent source of information for many topics, in particular for renewable energy sources such as wind and solar power. At the end of each section of the book, I have listed "references" that may be useful to readers. It is not possible to thank each author of these articles individually. Thus, I acknowledge their contributions here.

In the book I have also touched on topics not directly related to the design of cement plants, such as carbon capture and storage and cement substitutes such as Calera and Calix as these products could well be the future of green cement. I have also collected data for these topics from various sources.

ACC, KHD, ABG Cement, and Bhavya Cement gave me permission to include general layouts and departmental layouts of their cement plants in the book.

I have received valuable help from my friends, Mr. K.V. Pai, Mr. A.D. Deshpande, and Mr. P. Biksha Reddy.

Mr. V. Ramkumar helped me with flowcharts and layouts.

My publishers, M/S B.S. Publications, who are bringing out this, my third book, have done a great job. The production values are world class. I would like to thank Mr. Nikhil and Mr. Anil Shah of BS Publications.

I also would like to thank the executives and staff of BS Publications for their diligent efforts in bringing out the book in a short time.

I take great pleasure in thanking my grandsons, Pratik and Sanket Gupte, who helped me time and again with my errant computer system.

Last but not least, my wife, Urmila, and my daughter, Dr. Anuradha, have been my source of strength as always.

I hope and pray that my effort will be well received by the cement community.

As I am 80 years old, this will probably be my last book. I therefore wish all of my readers a very happy and prosperous future.

ABBREVIATIONS

SECTION 1

₹	Indian rupee
$	US dollar
£	British pound
€	euro
BFSC	blast furnace slag cement
CCNUCC	Convention Cadre des Nations Unies sur les Changements Climatiques
CCS	carbon capture and storage
CDM	Clean Development Mechanism
CII	confederation of Indian industries
CMA	Cement Manufacturers' Association
EAI	Energy Alternatives India
GHG	greenhouse gases
IEA	International Energy Agency
INCCA	Indian Network for Climate Change Assessment
kcal/kg	kilocalories per kilogram
kg	kilogram
kg/t	kilograms per ton
kwh	kilowatt hour
MoEF	Ministry of Environment and Forests
Natcem™	Natural Cement
NCCBM	National Council for Cement and Building Materials
NMEEE	National Mission for Enhanced Energy Efficiency
NO_x	oxides of nitrogen
OPC	ordinary Portland cement
PPC	Portland pozzolana cement
SO_x	oxides of sulphur
tpd	tons per day
tph	tons per hour
UNFCCC	United Nations Framework Convention on Climate Change
UNIDO	United Nations Industrial Development Organisation
VDZ	Verein Deutscher Zementwerke
VRM	vertical roller mill
WBCSD	World Business Council for Sustainable Development
WHR	waste heat recovery
WHRB	waste heat recovery boiler
WHRS	waste heat recovery system
WWF	World Wildlife Fund

SECTION 2

CEM II	eco-efficient cement
CEM II-S	OPC with blast furnace slag
CSN EN	European standards
EN 197-1	cement standards of Europe
PCA	Portland Cement Association
PCA R & D	Portland Cement Association Research and Development

SECTION 3

ECRA	European Cement Research Agency
IPCC	Intergovernmental Panel for Climate Change
MEA	monoethylamine

SECTION 4

AF	alternate fuels
AFR	alternate fuels and raw materials
ANZBP	Australia New Zealand Biosolids Partnership
AR	alternate raw materials
CC	Cement Company
CEMBUREAU	European Cement Association
CHNS-O analyzer	carbon, hydrogen, nitrogen, sulphur and oxygen analyzer
CPAG	Claudius Peters Group
CPCB	Central Pollution Control Board
FLS	FLSmidth
gcv	gross calorific value
GEPIL	Gujerat Enviro Protection & Infrastructure Ltd
HW	hazardous wastes
ISP	Intermediate Service Provider
KHD	Kloekener Humboldt Deutz
MSW	municipal solid wastes
mtpa	million tons per annum
PCB	Pollution Control Board
PCB	polychlorinated biphenyl
PCT	polychlorinated terphenyl
pH value	a measure of activity of hydrogen; pure water, $pH = 7$; $<7 =$ acidic; $>7 =$ alkaline
PREGA	Promotion of Renewable Energy &Greenhouse Gas Abatement
RDF	refuse-derived fuels
SPCB	State Pollution Control Board
UNDP	United Nations Development Programme

SECTION 5

DG set	diesel generator set
FOB	free on board
FOR	free on rail
KC	Kalina cycle
kcal	kilocalories
kcal/nm^3	kilocalories per normal cubic meter
KW/kw	1000 watts
MW/mw	1000 kilowatts
nm^3	normal cubic meter
OEC	Ormat energy converter
ORC	organic Rankine cycle
SRC	steam Rankine cycle
TPP	thermal power plant
TPS	thermal power station

SECTION 6

BEE	Bureau of Energy Efficiency
BIS	Bureau of Indian Standards
CFCs	chlorofluorocarbons
CoP	coefficient of performance
CSP	concentrated solar power
EER	energy efficiency ratio
EPA	Environmental Protection Agency
EPD	equipment power density
ETP	effluent treatment plant
FSC	Forest Stewardship Council
HCFCs	hydrochlrofluorocarbons
HDD	heating degree days
HVAC	heating, ventilation, and air conditioning
IEQ	indoor environmental quality
IGBC	India Green Business Centre
IPLV	integrated part load value
ISI	Indian Standards Institution
LCoE	levelized cost of energy
LEED	Leadership in Energy and Environmental Design
LEED-NC	LEED for new construction
LPD	lighting power density
NBC	National Building Code—India
NTP	normal temperature and pressure (0 °C)
O & M	operation and maintenance
PV	photovoltaic
RSPM	respirable suspended particulate matter

RWH	rain water harvesting
SHGC	solar heat gain coefficient
SPC	statistical process control
SPM	suspended particulate matter
STP	standard temperature and pressure (25 °C)
VOC	volatile organic compound

SECTION 7

ABB	Asean Brown Boveri
RP	roller press

SECTION 9

CSI	Computer Society of India
GPC	geo polymer cement

SECTION 10

CoP	cost of production

SECTION 1

Green Cement

Contents

CHAPTER 1

What is a Green Cement Plant

1.1. Definition of a green cement plant

The natural color of cement is gray, varying between lighter and darker shades. What then is green about a "green" cement plant?

Obviously, "green" does not refer to the color of the cement. It refers to the philosophy that lies behind the design concepts of new cement plants.

A green cement plant is one that is designed to conserve natural resources of all kinds and that contributes to the release of the greenhouse gases (GHG) to the atmosphere to the least possible extent consistent with the quality of cement produced.

1.2. Blended cements

Release of CO_2, a greenhouse gas, is inherent in the process of the manufacture of cement, as CO_2 is released from limestone, the basic raw material of cement during the process of calcining. One kilogram of calcium carbonate releases 0.44 kg of CO_2. Therefore, in making 1 kg of clinker, approximately 0.51 kg CO_2 gets released into the atmosphere.

The quantity is reduced when computed in terms of cements made from the clinker:

* OPC: $\sim$ 0.49-0.50 kg/kg
* PPC: $\sim$ 0.34-035 kg/kg
* BFSC: $\sim$ 0.19-0.20 kg/kg

This itself is a factor in reducing GHG emissions. Blended cements release less GHG as compared to OPC per ton of cement.

1.3. Combustion of fuel

A vital component of total carbon dioxide released to the atmosphere is the CO_2 released in the process of combustion of fuel fired in the kiln and calciner in the clinkerization process. The quantum released is directly related to the quantum of fuel fired and the quantum of carbon in it.

Designing Green Cement Plants
http://dx.doi.org/10.1016/B978-0-12-803420-0.00001-9
Copyright © 2016 BSP Books Pvt Ltd.
Published by Elsevier Inc.

Again, by the same logic, the obvious way to reduce emissions is to reduce the heat requirement, or what is called specific fuel consumption, and/or to use fuels with less carbon or those that are carbon neutral.

1.4. Alternate fuels

Alternate fuels have been successfully used in many countries in kilns and calciners. In Europe the cement industry is progressing toward zero fuel costs. Great possibilities exist for using wastes of industry and agriculture that have heat value as secondary fuels in kilns and calciners.

Certain modifications and additions are of course required in the existing fuel storing, preparation and firing systems in order to fire secondary fuels alongside of basic fossil fuels.

Whereas blended cements can be easily made in an existing plant, introduction of alternate fuels would require careful planning and engineering and also capital investment.

1.5. Electrical energy

Production of cement also requires a supply of electrical energy, expressed as kwh/ton of cement. Electrical energy is presently produced mostly by burning fossil fuels like coal and oil. Thus reduction of electrical energy by making cement indirectly means a reduction in electrical energy produced and thereby in GHG released to the atmosphere. If 1 kwh is used in a cement plant, the generating station has to produce much more to allow for transmission losses and for its own inputs. In some countries transmission losses are small, say 10%, but in some countries (India for one) they are more than 30%.

Any saving in electrical energy by the cement plant, howsoever small, has a still greater impact on decreasing GHG emissions when energy generated at thermal power stations is taken into account.

1.6. Waste heat recovery

Cement plants can further contribute significantly to reducing GHG emissions by converting waste heat in the exhaust gases from the kiln and cooler into electricity using waste heat recovery systems (WHRS). There is plenty of scope in existing dry process cement plants to produce power from waste heat.

Due to recent developments in technology it is now possible to generate power even from waste gases in modern cement plants with low heat contents by using the organic Rankine cycle and the Kalina process. It is estimated that between 20 to 30% of the energy required by a cement plant can be generated by installing WHRS. Energy so generated can be used in the plant or fed to the grid.

1.7. Renewable energy

All fossil fuels emit CO_2. Biomass fuels are carbon neutral. Sources of energy like wind, solar and hydraulic are not only totally free of carbon but on top are renewable, and also inexhaustible. Increasing attention is being paid to making them viable sources of energy.

Cement plants in various parts of the world are beginning to use wind and solar energy to meet part of their energy requirements.

1.8. Thus, making or designing a green cement plant in effect means:

1. providing facilities for making blended cements in an adequate measure
2. designing components like calciners to reduce obnoxious gases like NO_x, SO_2, etc.
3. provide providing for processing and firing alternative fuels which will reduce the quantum of CO_2 released
4. designing burners and firing systems for available alternative fuels
5. if required, providing for bypass of kiln gases which can contain excessive alkalis and chlorides as a result of firing certain alternative or waste fuels
6. providing for waste heat recovery systems to generate power or for other applications
7. consideration of making composite cements, which are a form of blended cements

1.8.1. To this list will soon be added:
1. using/making substitute cements
2. using renewable energy

1.9. Carbon capture

There are developments which aim at reducing the GHG emissions by physically collecting CO_2 emitted and storing it and making it available to other industries that have use for it, and even for making cements of new types.

1.10. Other aspects

Apart from the two major aspects described regarding sustainability and GHG emissions there is more to making a cement plant green:

1. keeping the environment green by planting trees and taking up schemes for afforestation
2. adopting more scientific mining methods that cause minimum damage to the environment by minimizing mining footprints
3. reclaiming used mines for landscaping, creating water reservoirs, etc.
4. creating green belts in and around plant and colony
5. installing water conservation schemes like rainwater harvesting, water treatment for recycling
6. designing and constructing green buildings in the cement plant wherever possible to make maximum use of natural light, ventilation, etc.

In short, the cement plant should blend beautifully with its surroundings.

1.11. Scope for making and designing green cement plants

The cement industry is consciously making efforts in various areas (listed in section 1.8) and is very much interested in making existing plants green and in designing new plants as green plants.

1.11.1. Blended cements

Presently almost all cement plants the world over are making blended cements. In India itself ~ 74% of cement made is blended cement. Slag adding up to 60% has been used up. Fly ash is available but further increase in the quantum of Portland Pozzolana Cement (PPC) is limited unless the ceiling to which fly ash can be added is raised. This change can only be sanctioned by national entities that govern standards of cement, like the Bureau of Indian Standards in India.

1.11.2. Alternate fuels

The main problem is the selection of a fuel that is steadily available in required quantities over a long period of time and which would have reasonably uniform physical and chemical properties, such as calorific value.

1.11.3. Waste heat recovery

Introducing waste heat recovery systems requires heavy capital investment and therefore requires careful planning and engineering.

In the subsequent sections and chapters all these aspects have been covered in detail so as to present a comprehensive picture of what it takes to make a green cement plant.

RECOMMENDED READING

1 Energy Efficiency & GHG emission reduction Initiatives by the Indian Cement Industry- by Dr. S. P. Ghosh, Cement Manufacturers Association 2008

2 Best practices & technologies for energy efficiency in Indian Cement Industry -by, Ashutosh Saxena, National Council for Cement & Building Materials

3 Output of a Seminar on Energy Conservation in Cement Industry- Handy Manual for Cement Industry by UNIDO 1994

4 Energy Efficiency improvement opportunities for Cement Industry by Ernst Worrell et al for Environment Energy Technology Division of Lawrence Berkeley National Laboratories, 2008

5 Energy Efficiency & PAT Programme by Dr. S. K. Handoo of Cement Manufacturers Association - Presentation at Green Cemtech 2011

6 Cement Sector- Estimation of Baseline & Target Setting by Chkaravarty for National Mission for Enhanced Energy Efficiency (NMEE).

CHAPTER 2

Greenhouse Gases

2.1 What are greenhouse gases?

Greenhouse gases are those gases that have a "greenhouse" effect; that is, they retain heat and have a warming effect, just as greenhouses that are used to maintain warm/temperate conditions to nurture plants.

2.2 Principal greenhouse gases are:

1. carbon dioxide (CO_2)
2. methane (CH_4)
3. oxides of nitrogen (NO_x)

Out of these, CO_2 and NO_x are the most relevant because they are an integral part of the process of production of cement. Carbon dioxide is released in the process of calcination and also in the process of combustion.

2.2.1. Carbon dioxide from the process of calcination

In the production of cement, the principal raw material, limestone, is calcined into calcium oxide, a major constituent of cement clinker. Portland Cement clinker contains about 65% CaO.

Limestone which is calcium carbonate in calcination releases carbon dioxide as per equation:

$$CaCO_3 = CaO + CO_2$$

$$100 = 56 + 44$$

For every kg of CaO, CO_2 released is 0.79 kg.

Therefore for every kg of Portland cement clinker, CO_2 released is ~ 0.51 kg.

(Intergovernmental Panel for Climate Change (IPCC) default value is 0.525 kg/kg clinker, corresponding to $\sim$67% CaO in clinker)[1]

---

[1] CII publication Low Carbon Road Map for Indian Cement Industry.

Designing Green Cement Plants
http://dx.doi.org/10.1016/B978-0-12-803420-0.00003-2

Expressed in relation to cement, it would be $\sim$ 0.49 kg per ton of

Ordinary Portland Cement (OPC) which has $\sim$ 5% gypsum in it, with a cement to clinker ratio 1.05:1.

In the case of blended cements, the ratio of cement to clinker is much higher.

In the case of Portland Pozzolana Cement (PPC), which can contain up to 30% fly ash, the cement to clinker ratio would be $\sim$ 1.54 (percentage of gypsum remaining the same).

Therefore CO_2 released per kg of cement would be $\sim$ 0.33 kg. In the case of slag cement, which can contain up to 60% blast furnace slag, the cement to clinker ratio would be 2.86 and the CO_2 released per kg of cement would be $\sim$ 0.19 kg.

Thus the obvious way to reduce emissions of CO_2 in relation to cement produced is to make blended cements.

2.2.2. Carbon dioxide from the process of combustion of fuel

In producing cement clinker, heat is first required to calcine raw meal fed to the kiln. Raw meal is then **sintered** at a temperature of approximately 1400 °C to produce cement clinker.

Presently most of the cement plants are dry process cement plants with 5/6 stage preheaters and calciners and high-efficiency clinker coolers.

The heat is supplied by firing fuel, mostly coal, sometimes oil or gas.

To derive maximum benefit of the calciner, fuel is divided between the kiln and calciner in the ratio 40:60.

Total heat consumed in making clinker is expressed as sp. fuel consumption in kcal/kg clinker. It takes into account various losses in the process.

Presently the representative value of sp. fuel consumption can be taken as 700 kcal/kg, though quite a few plants have achieved efficiency of 650 kcal/kg.

Coal used in making cement varies greatly in quality, depending on its source. The calorific value of coal is around 4500–5000 kcal/kg.

Coal often comes from different sources and therefore varies in calorific value, ash content, and moisture.

Most cement plants therefore install facilities for blending coal in the form of circular/linear stacker reclaimers so that they can fire in kilns and calciners coal of uniform quality.

Let us assume for the sake of this exercise that the representative value coal is:

> useful calorific value: 4500 kcal/kg
> ash < 30%
> fixed carbon ∼ 50%

For a supply of 700 kcalories, coal used would be $700/4500 =$ 15.6% or 0.156 kg per kg clinker. It would contain 50% or 0.078 kg carbon, which will produce carbon dioxide:

$$C + O_2 = CO_2$$

$$12 + 32 = 44$$

Therefore the coal burnt would produce 0.29 kg of CO_2 per kg clinker.

2.2.3. Total impact

Adding this to the CO_2 released by way of calcination, total CO_2 released $= 0.51 + 0.29 = 0.80$ kg/kg clinker emission in terms of kg/kg cement would be:

		Cement/clinker ratio	CO_2 released kg/kg cement
1	OPC	1.05/1	∼ 0.76
2	PPC (30% fly ash)	1.54/1	∼ 0.52
3	BFSC (60% slag)	2.86	∼ 0.28

The beneficial impact on reduction of GHG emissions as result of making blended cements can clearly be seen from the above table.

2.3 NO_x

It is possible to control generation of NO_x by admitting feeds of raw meal, fuel, and tertiary air at multiple levels in the calciner to create reducing zones. A reducing agent with an ammonia content of about 5% can also be added in the reducing zone to bring down NO_x concentration.

2.4 Potential for Reduction of Current GHG Emissions when making Blended Cements

Potential for reduction in current GHG emissions by making blended cements will differ from country to country.

The principal factors influencing it would be:

1. Proportion of blended cements already being made in relation to total volume of production.
2. Division between PPC and BFSC – is the proportions of these two types of cements made in the total quantity of blended cements produced.
3. The three industries involved—cement, steel, and power. Suppliers of blending materials are the power and steel industries.

   Growths of these three industries may or may not synchronize and hence slag or fly ash may not be always available in quantities required by the cement industry.

   There is a limit to the proportion of fly ash that can be added.

   Today the maximum permitted is 35%.

   Unless this proportion can be increased there is little scope for increasing the quantum of PPC produced.

4. In India, presently the proportions of different types of cement made are[2]:
   OPC 25%
   PPC 66%
   BFSC 8%

   All available slag is used. At the current production level, therefore, the maximum scope is to make all OPC into PPC.

---

[2] CII publication Low Carbon Road Map for Indian Cement Industry & CMA publication Cement Statistics 2011.

Further, if the proportion of PPC averages 27% it can be raised to the maximum permissible 35%.

The potential for reduction can be worked out.

Since the situations will be different from country to country, with standards prevailing also playing an important role, each country will have to work out its own potential for reduction of GHG emissions by making blended cements.

5. Fortunately no major changes are involved in beginning to make blended cements or enhancing the proportion thereof.

What is required is storage for the blending materials—fly ash or slag, and their withdrawal and feeding to cement mills.

This is easily possible in most cases in existing plants and can be planned in new plants.

In new plants there is also a possibility of choosing grinding systems with vertical roller mills and high-efficiency separators or a roller press and separators so that energy consumption can also be brought down.

Since both fly ash and slag are processed materials and are also waste materials they are available cheaply. Therefore making blended cements is also a highly profitable proposition.

Further details are covered in a separate chapter.

The cost of blast furnace slag varies at around Rs. 850 ($15.5) per ton.

The cost of fly ash is around Rs. 750 ($13.6) per ton.

2.5 Potential for reduction in GHG emissions due to combustion of fuel

1. The present level of sp. fuel consumption in dry process plants is ~ 700 kcal/kg. In the foreseeable future it may be brought down to 650 kcal/kg. Emission of GHG on account of combustion of fossil fuel is thus likely to decrease from 0.29 kg to 0.27 kg/kg clinker.
2. If, however, fossil fuels are replaced by other fuels collectively called secondary or alternate fuels with less carbon or by fuels which are carbon neutral there is greater scope for reduction of GHG emissions.

For example, if a carbon neutral fuel is used to the extent of 20%, resultant carbon would be 40% and GHG emission at 700 kcal/kg would be 0.23 kg/kg clinker.

Alternate fuels thus assume great importance on three counts.

1. They can help bring down GHG emissions.
2. They reduce the problem of disposal of wastes (the majority of alternative fuels are industrial or agricultural wastes.
3. They help conserve fossil fuels. Further details are covered in a separate chapter.

2.6 Potential for reduction of GHG emission on account of waste heat recovery

As mentioned in the previous chapter, installing waste heat recovery systems (WHRS) is another way to reduce net GHG emissions.

CO_2 contained in the waste gases used to generate power by WHRS is already accounted for.

Electrical energy is universally bought from centralized power plants, mostly thermal and hydraulic. Many cement plants also have captive power plants which can supply up to 40% of their requirement. In hydraulic power plants there are no polluting gas emissions but the quantum of electrical energy produced is generally a small proportion of the total produced, except in hilly areas where all of the power generated may be from hydraulic power plants.

Thermal power plants are the major source of power for industry and all other sectors.

Thermal power plants use coal for fuel combustion, which produces exhaust gases containing a good amount of CO_2. Kilocalories burnt to produce one unit of electric energy kwh are the yardstick for measuring performance of thermal power plants.

For an overall efficiency of 40%, the kcal/kwh would be 2150.

Coal supplied to power stations is generally inferior to that supplied to cement plants and contains + 35% ash. In India the calorific value is between 3500–4000 kcal/kg coal.

Using the same procedure, and assuming a calorific value of coal of 3500 kcal/kg with 45% carbon, CO_2 emitted per kwh works out to ~ 1 kg for every kwh in a cement plant. Let us say 2 kwh are from the Thermal Power Station (TPS), to allow for transmission and other losses. By not buying 16 units from the TPS, the cement plant saves actually 32 units from the TPS. This results in saving 32 kg of CO_2 per ton of cement (OPC), or 0.032 kg/kg.

Thus, indirectly through WHRS, cement plants can achieve significant reduction in GHG emissions as well as conservation of fossil fuels.

2.7 Composite and low carbon cements

These are variations of blended cements that would result in reduced GHG emssions. There will be more about them in the chapter on blended cements.

2.8 Carbon Capture

As mentioned earlier, this strives to separate the CO_2 in exhaust gases and store it in a safe place. It would be available to prospective buyers who need it in their processes of manufacture. There will be more about this later.

2.9 Renewable sources of energy

Power generated from fossil fuels will produce carbon dioxide. Sources of power like the wind, sun and sea would be totally free of GHG emissions. Hence in the future there will be greater research directed to making power from these sources commercially viable. Some cement companies have already set up wind farms and solar energy farms to produce electricity from the wind and sun. Their number is increasing.

2.10 Green cement plants in the immediate future will thus have the following main features with regard to manufacturing processes and activities. They would:

1. be geared to make blended and composite cements
2. use alternate fuels
3. produce power from WHR
4. possibly set up alternative sources of renewable energy

2.11 Other options

Research is also underway to find substitutes for cement as we know it. Substitutes would also tend to reduce GHG emissions.

Among the various possibilities there are:

1. Calera process.

It makes cement by reaction of carbon dioxide with minerals in sea water. Thus carbon dioxide released by thermal power plants and cement plants is raw material for the Calera Process.

2. Novacem

It is based on using magnesium silicate rather than limestone.

3. Calix

This makes use of dolomitic limestone.

4. Geopolymer Cements

They are made from waste materials like fly ash and slag and concrete wastes by activation with alkalis.

5. Aether

It is claimed that the loss of ignition in raw mix for this cement is $\sim 30\%$, as opposed to 35% for OPC raw mix.

Further, sintering temperatures are 1250–1300 °C, compared to 1350–1400 °C for OPC clinker.

RECOMMENDED READING

1 Emission Reduction of GHG from Cement Industry by C A Hendricks et al for GHG R&D Programme of IEA, 2004

2 CO_2 Emission from Cement Production by M J Gibbs et al for Good Practice & Uncertainty Management in National GHG Inventories

3 How to turn around trend of cement related emissions in Developing World by Nicolas Muller et al - Report for WWF International

4 Towards a Sustainable Cement Industry- Independent study commissioned by World Business Council for Sustainable Development (WBCSD)- a Batelle Report 2002

5 CDM opportunities in Indian Cement Industry- report by Ernst & Young for Confederation of Indian Industries (CII)- 2009

6 Cement Sector GHG emission reduction- case studies- a Consultant's report by Loreti Group for California Energy Commission 2009

7   Environmental impact of Thermal Power Generation - ENZEN Global Solutions

8   Fuel & GHG emission reduction by fuel switching and technology improvement in Iranian Electricity Generation Sector- International Journal of Engineering -vol 3, issue 2

9   GHG Reduction option for Indian Cement Industry by Dr. Jyoti Parekh et al for Integrated Research & Action for Development 2009

10 Reduction of GHG from cement production- through combustion optimization in Romania - by Cimus- Holcim-Cimpulung, Romania

11 India GHG emissions - 2007 , by Indian Network for Climate Change Assessment, INCCA- Ministry of Environment & Forests

12 Emission profile 2007, Cement Process & Fuel Combustion in cement manufacture by Dr. S.P.Ghosh of CMA for INCCA

13 Cement Technology Road Map 2009 - Carbon Emissions Reduction upto 2050 (WBCSD & IEA)

14 Contribution of Electric Power Technology to GHG reduction by Kanik Urashima et al for Science & Technology Trends- Quarterly Review 2009

15 Uncertainty reduction in GHG emissions by NatCom India, for MoEF

16 Energy use efficiency in Indian Cement Industry- Application & Data Enveloping Analysis by Saroj Mandal- working paper for Institute for Social & Economic Change Bangalore.

17 Baseline Methodology for GHG reduction for WHR & Utilization for Power Generation in Cement Plants- CDM Executive Board- UNFCCC/CCNUCC- 2008

18 VDZ Activity Report- Environmental Protection in cement manufacture 2003-05.

CHAPTER 3

Summary

3.1 Reduction of emissions of greenhouse gases

The well-established route to reducing GHG emissions arising from the process of calcination is to make blended and composite cements.

Other possibilities include

1. capturing carbon dioxide from exhaust gases and storing it, and
2. developing substitutes which would replace cement as we know it today.

3.2 Blended cements

It is easy to make blended cements. Even existing plants can easily add facilities to bring in slag and fly ash to make blended cements.

Composite cements can similarly be made in both new and existing plants.

3.2.1 The extent to which the quantity of blended cements can be increased depends on:

1. the extent to which they are already being made
2. the availability of blending materials: slag and fly ash in the quantities required
3. the possibility of increasing the permissible quantities of these materials in blended cements. This decision depends primarily on users of cement and also on approval by the standards institution that establishes quality and performance standards for cements in each country.

This is a long, drawn-out process and will have to be expedited.

3.2.2 The same is the case for composite cements. A couple of decades ago even PPC and BFSC were considered adulterated cements. There will be similar questions about the suitability of composite cements.

Designing Green Cement Plants
http://dx.doi.org/10.1016/B978-0-12-803420-0.00004-4

3.3 Carbon capture and storage

Considerable attention is being given to this method for reducing GHG emissions as it will drastically bring down emissions both from the calcination process and from combustion not only in cement plants but also in thermal power plants.

While the technology for treating captured CO_2, compressing it and transporting it in liquid form in pipelines is available, it has not yet been applied on the large scale required in the case of the power and cement industries.

Storage of CO_2 in large quantities either underground or deep under the sea requires development of such storage in excavated mines, oil reservoirs, geological rock formations, etc. The first step would be to locate likely areas.

The whole stage of transporting CO_2 and storing it is energy intensive and also has social and political dimensions.

It is unlikely that CCS will become a reality to a sizable extent in the cement industry before 2020. But it is a possibility.

3.4 Cement substitutes

It is time that alternatives to OPC and its variants are available for construction. Concentrated and aggressive research in this direction is called for.

There are two main issues:

1. that the substitute(s) should be available on the scale presently occupied by OPC and other variants
2. that it is possible to make the substitutes in the plants actually producing cements. Fresh investment to produce substitutes on a scale to suit the existing level of consumption would be commercially unviable if present cement-making capacity is rendered idle.

Even more important is the acceptability of the substitutes by users and by standards institutions and a host of other involved parties.

However, since some substitutes use CO_2 released by thermal power plants to make cement, there is a distinct possibility of cement plants making new types of cement by using CO_2 from kiln exhaust gases and from captive power plants.

3.5 Therefore at this stage to design green cement plants the cement industry will have to depend on:

1. making blended and composite cements
2. using alternate fuels
3. installing waste heat recovery systems.

3.6 Renewable energy sources

Renewable energy (RE), besides conserving fossil fuels, is also free of GHG emissions and would thus make a significant contribution to reduction in GHG emissions.

The problems to be overcome with the use of RE appear to be:

1. Its inconsistency in terms of availability and quantity.

   Winds change direction, velocities keep changing, and hours are irregular.

   Solar energy is generated during daytime. The amount of sunshine and its intensity are variable.
2. Capital costs of investment expressed as rs/$/€ per mw are high.
3. Large areas of land need to be acquired to set up wind and solar farms.
4. Sites for cement plants may not always be suitable for wind farms.
5. Costs for RE production should be comparable to costs for grid power.

These problems are being tackled on a priority basis, and there are signs that they will be overcome in the near future.

RECOMMENDED READING

1 95th Report on Performance of Indian Cement Industry - presented to Parliament of India in 2011
2 Sustainable Cement Industry by Sanghi & Bhargava at Workshop on International Comparison of Industrial Energy Efficiency 2010
3 Technological Trends in Cement Industry- Energy & Environmental Impact by G. Jayaraman
4 Record Growth & Modernization in Indian Cement Industry- NCCBM News Letter.

SECTION 2

Blended Cements

Contents

List of Figures

Chapter 1

Chapter 2

List of Tables

Chapter 1

CHAPTER 1

Blended Cements and Designing Cement Plants to Make Blended Cements

1.1 The impact of blended cements on reducing GHG emissions into atmosphere has been described in Chapter 2 of Section 1.

1.1.1 Cement/clinker ratio

Blended cements increase the cement/clinker ratio. GHG emissions are reduced per ton of cement produced, other things remaining the same.

For example:

Type of cement	Cement/ clinker ratio	Total GHG emissions kg/kg cement	% Reduction
OPC	1.05	0.77 median value	–
PPC (30% fly ash)	1.54	0.52	32
BFSC (60% slag)	2.86	0.28	63

1.2 Win-win situation as a result of making blended cements

The most effective way to reduce GHG emissions would be to make blended cements.

Since both fly ash and slag are waste products of their respective industries, cost of their procurement is far less than the cost of production of the clinker that they would replace.

The cost of blast furnace slag is around Rs. 850/ton ($15.5) and that of fly ash is around Rs. 750/ton ($13.6).

This is thus a win-win situation for all because:

1. The power and steel industries get rid of their respective waste products.

Designing Green Cement Plants
http://dx.doi.org/10.1016/B978-0-12-803420-0.00002-0

2. The cement industry gets a useful material at much less cost.

3. The community benefits as the waste products are not dumped in landfills, thereby saving large areas of land.

4. There is significant contribution to reduced global warming effect by reduced GHG emissions.

1.3 Potential for increasing production of blended cements

Blended cements are already being made to a large extent in most countries of the world. Their proportion to the total cement produced varies from country to country.

Also, proportions of PPC and BFSC in the quantum of blended cement produced will vary from country to country.

Further potential for reduction of GHG emissions would therefore vary from country to country.

1.4 Availability of fly ash and slag

A major factor in increasing the quantum of blended cements produced is the availability of the blending materials, fly ash, and slag.

Though at present permissible additions of fly ash and slag are 30% and 60% respectively, it is very likely that these values will go up in the near future.

Production of cement has been increasing at a very rapid rate in countries like India and China. To meet the increasing demand of construction and industrial activities, both the steel and power sectors will also be expanding.

However, the three sectors, namely cement, power, and steel, may not have synchronized growths. Therefore in some countries at some time one or the other of the blending materials may not be available in sufficient quantities, putting brakes on the production of blended cements.

1.5 Logistics of transport of fly ash and slag

Logistics of the transport of fly ash and slag to cement plants will play an important part in the rate of growth of blended cements. Cement plants get located near limestone deposits; steel plants near iron ore deposits. Thermal power stations tend to get located near coal deposits.

It is not always possible for the three to be located close to one another.

Therefore fly ash and slag have to be transported to cement plants. This creates some limitations and introduces aspects of storage and handling of these materials in cement plants.

When market demands, clinker is transported to a steel plant to make BFSC.

1.6 Increasing quantum of permissible additions of fly ash and slag

Yet another possibility is to increase the permissible quantities of these blending materials in the manufacture of PPC and BFSC. Here three parties are involved.

Firstly the **consumers,** that is the construction industry, must accept further increase in the quantity of blending materials. They must be convinced that it will not result in deterioration in the quality of cement or construction. The most important properties with regard to early and final strengths should not be adversely affected.

It takes a long time to remove consumer bias. Two decades back PPC and BFSC had no market as they were considered adulterated cements.

Now that blending has yet another perspective, that of reducing GHGs, it is necessary for such biases to be removed.

1.6.1 The second party that can help with promotion is constituted by **research institutes** the world over. They should, by conducting urgent and serious research, examine all aspects of the increase in quantities of these proven blending materials and satisfy themselves and the cement makers and cement users that there are no disadvantages whatever.

1.6.2 The third party includes worldwide **standards institutions**, whose responsibility it is to lay down standards for the qualities of cements produced. They also have the responsibility to ensure that cements produced conform to the prevailing standards.

Unless the standards institutions endorse the revisions and suitably revise the respective cement standards, the acceptance of cements made with higher percentages of blending materials will be slow.

1.7 Urgent review required

This matter has therefore to be viewed in a wider perspective and all parties concerned should come together to agree and to formulate a policy. In India, for example, bodies like the CMA (representing cement manufactures), NCCBM (representing the apex research institute), BIS (representing the standards institution) and builders' associations (representing users) should evolve and speedily standardize the quality of blended cements to be produced in the future.

1.8 Making of blended cements

There is nothing new in making blended cements, as both PPC and BFSC have been made the world over for decades.

The processes of making PPC and BFSC are well known, as is the machinery used.

There are of course small variations, such as:

1. Whether to grind clinker and fly ash/slag together or separately.
2. Whether fly ash should be introduced first in the mill or in the separator.
3. Selection of the grinding system—tube mill, vertical roller mill or roller press and ball mill.

They can be examined technically on a case-by-case basis to arrive at optimum solutions for specific projects.

1.9 Differences on account of increase in size of cement plants

1. Cement plants have grown much bigger in size. Representative size of plant is about 3-8 mtpa in terms of cement produced in one place.
2. The storage and handling facilities for these materials have to be designed from this angle.

   This will require larger quantities of fly ash and slag to be brought in and stored. Slag may have to be dried as well.

3. Fly ash will be stored in silos that are larger in numbers, size and capacity. Fly ash will be conveyed to mills mostly pneumatically.

Slag will be stored in covered gantries with facilities for receiving wet slag and also for dried slag. A drier will be a necessity if slag is received wet.

If slag is ground separately, it can be ground in the Vertical Roller Mill (VRM).

4. While fly ash will be brought in special self-unloading carriers by road in most cases, on occasions it may have to be brought in by rail in wagons similar in design to those used to transport cement in bulk.

1.10 Receipt of slag

Because of its quantity, slag will come in rake loads by wagons. The number of rakes to be handled being quite large, a large railway siding with wagon tipplers will be necessary to cope with the volume of slag to be received and unloaded.

To keep down costs of investment, quantities to be stored could be reduced consistent with maintaining continuity of production.

1.11 Impact on plant layout

A corollary of making blended cements is that the quantity of OPC made would be reduced correspondingly. More blended cements will be stored than OPC.

One mill and one silo may be reserved for OPC, with the rest for whatever type(s) of cement (PPC/BFSC) are to be made.

If a cement plant is making both PPC and BFSC (which is quite possible) groups of mills will be reserved for each category so as to avoid changes of settings, cleaning of silos to avoid contamination, etc.

1.11.1 Dispatches

In a great number of countries cements are mainly dispatched in bulk. Countries like India are following suit. Layout of the cement dispatching section should be designed to meet the requirements of bulk and bagged cements.

Further dispatches by road will predominate over dispatches by rail.

Layouts of packing plants will be designed to facilitate them.

1.11.2 Railway siding

Plant capacities at a given place have doubled or even trebled at some places, and the volume of traffic has increased. Because of the shift to dispatches by road, railway sidings within the plant are more likely to be designed for receiving coal and slag than for handling cement.

1.12 Various aspects of handling fly ash and slag have been dealt with in detail in Chapters 28 and 29 of Section 6 of the author's "Handbook for Designing Cement Plants." They are relevant now also.

1.13 Choice of grinding mill

Presently VRMs are replacing tube mills for grinding slag and even clinker. Thus there are small differences in layouts of cement mills to allow for the use of VRMs and corresponding auxiliaries.

The principal difference is the point of feeding fly ash in VRM systems. VRMs have built-in high-efficiency separators. Fly ash, being a naturally fine product, can bypass the mill and be fed to the separator at the mill.

The roller press and ball mill combination is also being adopted in a variety of ways to make cement.

1.14 System flow charts and layouts

Flow charts Figs. 2.1.1–2.1.3 show grinding circuits for OPC and blended cements based on the VRM as the grinding mill.

Figs. 2.1.4 and 2.1.5 show schematic layouts for making PPC and BFSC respectively, using VRMs.

1.15 Sectional capacities and storages to be provided

Table 2.1.1 shows the capacities of cement mills for clinkering capacities of 5000, 7500, and 10,000 tpd when making OPC, PPC, and BFSC.

1.16 Storages to be provided

Table 2.1.2 shows the requirements of various materials like clinker, gypsum and blending materials like fly ash and slag per day for different capacities and quantities to be stored along with clinker.

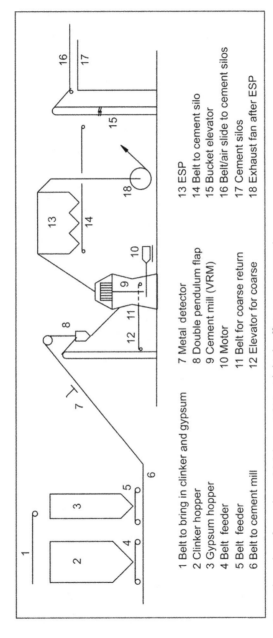

1 Belt to bring in clinker and gypsum
2 Clinker hopper
3 Gypsum hopper
4 Belt feeder
5 Belt feeder
6 Belt to cement mill

7 Metal detector
8 Double pendulum flap
9 Cement mill (VRM)
10 Motor
11 Belt for coarse return
12 Elevator for coarse

13 ESP
14 Belt to cement silo
15 Bucket elevator
16 Belt/air slide to cement silos
17 Cement silos
18 Exhaust fan after ESP

Figure 2.1.1 Flow chart for making OPC (using VRM with high-efficiency separator).

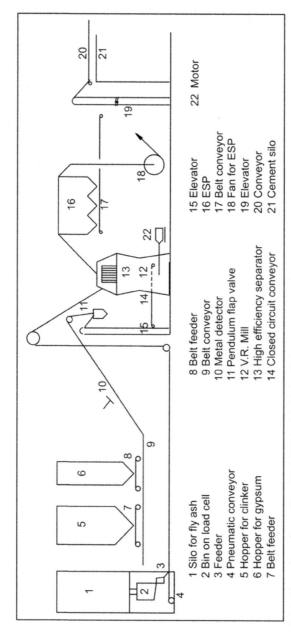

1 Silo for fly ash
2 Bin on load cell
3 Feeder
4 Pneumatic conveyor
5 Hopper for clinker
6 Hopper for gypsum
7 Belt feeder

8 Belt feeder
9 Belt conveyor
10 Metal detector
11 Pendulum flap valve
12 V.R. Mill
13 High efficiency separator
14 Closed circuit conveyor

15 Elevator
16 ESP
17 Belt conveyor
18 Fan for ESP
19 Elevator
20 Conveyor
21 Cement silo

22 Motor

Figure 2.1.2 Flow chart for making PPC (VRM with high-efficiency separator).

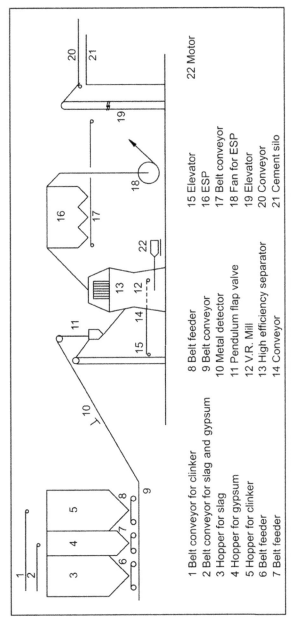

1 Belt conveyor for clinker
2 Belt conveyor for slag and gypsum
3 Hopper for slag
4 Hopper for gypsum
5 Hopper for clinker
6 Belt feeder
7 Belt feeder

8 Belt feeder
9 Belt conveyor
10 Metal detector
11 Pendulum flap valve
12 V.R. Mill
13 High efficiency separator
14 Conveyor

15 Elevator
16 ESP
17 Belt conveyor
18 Fan for ESP
19 Elevator
20 Conveyor
21 Cement silo

22 Motor

Figure 2.1.3 Flow chart for making BFSC (slag cement) (VRM with high-efficiency separator).

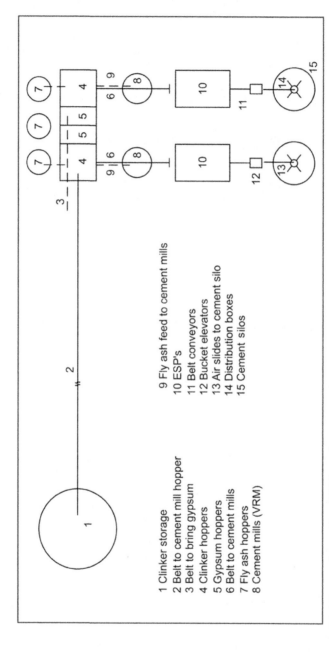

1 Clinker storage
2 Belt to cement mill hopper
3 Belt to bring gypsum
4 Clinker hoppers
5 Gypsum hoppers
6 Belt to cement mills
7 Fly ash hoppers
8 Cement mills (VRM)

9 Fly ash feed to cement mills
10 ESP's
11 Belt conveyors
12 Bucket elevators
13 Air slides to cement silo
14 Distribution boxes
15 Cement silos

Figure 2.1.4 Layout for making PPC (pozollana cement).

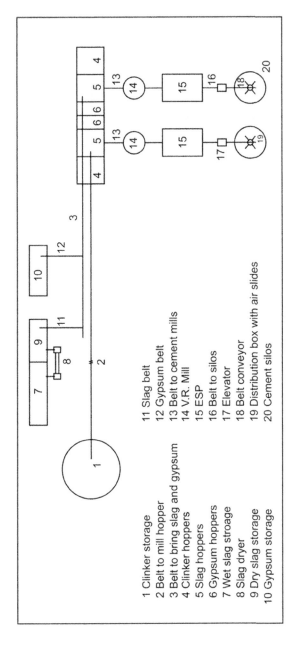

1 Clinker storage
2 Belt to mill hopper
3 Belt to bring slag and gypsum
4 Clinker hoppers
5 Slag hoppers
6 Gypsum hoppers
7 Wet slag stroage
8 Slag dryer
9 Dry slag storage
10 Gypsum storage

11 Slag belt
12 Gypsum belt
13 Belt to cement mills
14 V.R. Mill
15 ESP
16 Belt to silos
17 Elevator
18 Belt conveyor
19 Distribution box with air slides
20 Cement silos

Figure 2.1.5 Layout for making BFSC (slag cement).

Table 2.1.1 Capacities of Cement Mills Making OPC and Blended Cements

Cement Grinding	Cement Mill	Sizing Factor	Hours/Day	Multiplying Factor
	OPC	1.27	20	0.064
	PPC 30% ash	1.86	20	0.093
	BFSC 60% slag	3.46	20	0.173
		Kiln capacity tpd		
		5000	7500	10,000
	Mill making	**Mill capacity in tph**		
	100% OPC	320	480	640
	100% PPC	465	700	930
	100% BFSC	865	1300	1730

It also shows stocks of cements to be maintained (as per convention) for OPC, PPC, and BFSC, assuming 100% production of that type in each case.

1.17 General cement plant layout

Typical general layouts of a cement plant with a clinkering capacity of 10,000 tpd and making 6.5 mtpa of OPC and BFSC together were developed to illustrate the various aspects mentioned above regarding the design of layout for cement grinding and dispatching sections. These are included in Chapter 5 of Section 7 (Figs. 7.5.1 and 7.5.2).

Table 2.1.2 Storages of Clinker, Slag, Fly Ash and Cements at ...

Sr. No.	Material	Parameter	Unit	Clinkering Capacity (tpd)		
				5000	7500	10,000
1	Clinker	Factor	1.1	1.1	1.1	1.1
		Daily req.	tons	5500	8250	11,000
		# days stock Storage	14	14	14	14
			tons	**77,000**	**115,500**	**154,000**
2	Gypsum	Factor	0.07	0.07	0.07	0.07
		Daily req.	tons	350	525	700
		# days stock Storage	30	30	30	30
			tons	**10,500**	**15,750**	**21,000**
3	Fly ash	Factor	0.4	0.4	0.4	0.4
		Daily req.	tons	2000	3000	4000
		# days stock Storage	3	3	3	3
			tons	**6000**	**9000**	**12,000**
4	Slag	Factor	0.85	0.85	0.85	0.85
		Daily req.	tons	4250	6375	8500
		# days stock Storage	2	2	2	2
			tons	**8500**	**12,750**	**17,000**
5	Cement OPC	Factor	1.27	1.27	1.27	1.27
		Daily req.	tons	6350	9525	12,700
		# days stock Storage	7	7	7	7
			tons	**44,450**	**66,675**	**88,900**
6	Cement PPC	Factor	1.82	1.82	1.82	1.82
		Daily req.	tons	9100	13,650	18,200
		# days stock Storage	3	3	3	3
			tons	**27,300**	**40,950**	**54,600**
7	Cement BFSC	Factor	3.18	3.18	3.18	3.18
		Daily req.	tons	15,900	23,850	31,800
		# days stock Storage	3	3	3	3
			tons	**47,700**	**71,550**	**95,400**

– Number of

CHAPTER 2

Composite Cements

2.1 Composite Cements

Blended cements with more than one blending material are called composite cements.

For instance:

Clinker + fly ash + blast furnace slag
Clinker + fly ash + limestone powder
Clinker + blast furnace slag + limestone

2.2 Slag and fly ash complement each other in a composite cement. Composite cement made with 60% OPC + 30% fly ash + 5% limestone powder has higher strength than PPC made with 30% fly ash.

2.3 Production of multi-component cements enables not only fuel energy savings (by 30-40%) but also increased volumes of concrete production.

Composite cements with fly ash additive can be used to produce concretes for special application. They are suitable for producing alkaline and sulfate corrosion-resistant concrete.

Use of fly ash-slag mix allows optimization of the main characteristics of cement clinker and reduction of CO_2 emissions due to a greater cement/clinker ratio.

2.4 In the future, emitting CO_2 may attract penalties. Therefore clinker will be increasingly replaced by materials like fly ash, slag, limestone powder, natural pozzolanas, etc.

2.5 The type of additive used for making composite cement will depend on its availability locally. From this point of view, limestone powder suggests itself as the most convenient additive for use in cement plants.

Designing Green Cement Plants
http://dx.doi.org/10.1016/B978-0-12-803420-0.00006-8

2.6 The European Cement Standard EN 197-1[1] uses Portland Composite Cement as a generic term for the entire Group of CEM II cements. They include CEM II-S Portland slag cements. This category includes:

Portland—silica flume cement
Portland—pozzolana cement
Portland—fly ash cement
Portland—burnt shale cement
Portland—limestone cement

It is also used for CEM II M cements. They are cements containing Portland cement clinker plus other constituents like slag and silica flume that can be combined with one another.

These cements, in addition to the prospects for CO_2 reduction and conservation of resources, also offer considerable opportunities for optimizing properties like workability, strength development, and durability.

2.6.1 Portland composite cements CEM-II M made with granulated slag and limestone have been primarily used so far for industrial purposes.

Similar results can be expected from CEM-II S cements made from fly ash and limestone complying with EN 450 as main cement constituents.

2.7 Making of composite cements

Plants that are already making blended cements can easily and conveniently make composite cements.

For example, those that are making PPC or BFSC can add a hopper and feeder for limestone powder. Those that are making PPC can arrange to add slag, and vice versa.

2.8 The proportions of the third component to be added, whether slag, fly ash or limestone, have to be decided in consultation with research institutes and in accordance with prevailing (or forthcoming) standards for composite cements.

2.9 Flow chart Fig. 2.2.1 shows modifications required to convert an existing facility producing blended cement into one making composite cement.

---

[1] EN 197-1: Cement Standards of Europe; CEM II: Eco Efficient Cements; CEM II-S: OPC with blast furnace slag.

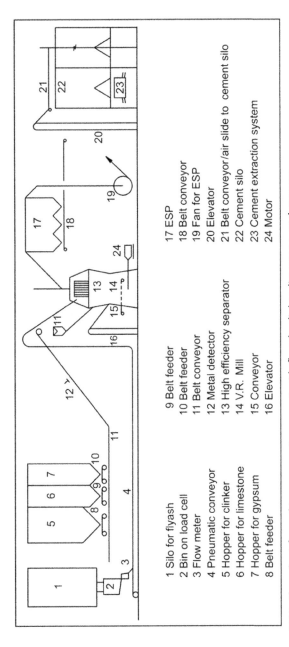

1 Silo for flyash
2 Bin on load cell
3 Flow meter
4 Pneumatic conveyor
5 Hopper for clinker
6 Hopper for limestone
7 Hopper for gypsum
8 Belt feeder

9 Belt feeder
10 Belt feeder
11 Belt conveyor
12 Metal detector
13 High efficiency separator
14 V.R. Mill
15 Conveyor
16 Elevator

17 ESP
18 Belt conveyor
19 Fan for ESP
20 Elevator
21 Belt conveyor/air slide to cement silo
22 Cement silo
23 Cement extraction system
24 Motor

Figure 2.2.1 Flow chart for making composite cement with fly ash, clinker, limestone, and gypsum.

2.10 Of necessity in a cement plant that simultaneously produces different types of cements arrangements would have to be made to store them separately.

Configuration of mills making these cements and corresponding silos for storage would have to be worked out carefully, taking into account the volume and consistency of demand for each type of cement produced.

2.10.1 For small and infrequent demands a multi-compartment silo could be used to store different types of cement.

Again, factors like:

dispatches by road and or rail and their respective volumes

dispatches bagged or in bulk

would have to be taken into account in planning the overall layout of cement grinding, storage, and dispatch sections.

Thus a layout would have to be worked out specifically for each plant to meet its requirements as well as possible.

RECOMMENDED READING

1 Low energy consuming modified composite cements and their properties by Sanylsky M & others Chemistry & Chemical Technology-vol 5 no 2, 2011
2 Fly ash -limestone ternary composite cements-synergetic effect at 28 days. by Klartje Dee Weerdt & others
3 Fly ash Blended Cements in India - present & future
4 Performance of Portland Composite Cements by Dr.Ing. Ch Muuller Cement International 2/2006
5 New Cement Specifications & Cement Types NCCBM Road Pavement Forum 2001

SECTION 3

Carbon Capture and Storage Systems (CCS)

Contents

List of Flow Charts and Plates

CHAPTER 1

Carbon Capture and Storage (CCS)

1.1 Limitations of reducing GHG emissions

Reduction of CO_2 in exhaust gases by employing any one or all of the following processes has inherent limitations.

Blended cements:

Limits prescribed by authorities cannot be exceeded until respective cement standards are revised.

Already almost all countries are making mostly blended cements. There is not much scope to increase the quantum further.

Improving fuel efficiency:

Considering the present state of the art of cement making, there is very little scope for reducing fuel consumption further (from 700 kcals to 650 kcal/kg), unless clinkerization takes place at lower temperatures.

Use of secondary/alternate/carbon neutral fuels:

Many a country has yet to reach the full potential of substituting fossil fuels with alternate/carbon neutral fuels, although European countries seem to have reached the zero fuel costs level.

But alternate fuels can replace fossil fuel on the whole by about 20-40%.

Waste heat recovery (WHR):

Developed countries have made good progress in this direction. Major producers like India have a long way to go. The potential for generating power by installing WHR is shrinking because of improvements in fuel efficiency even when state of the art technologies like the Ormat Rankine cycle or Kalina cycles are adopted.

Approximately 20% of plant requirements can be met by installing WHR systems under the circumstances.

Designing Green Cement Plants
http://dx.doi.org/10.1016/B978-0-12-803420-0.00007-X
49

1.2 Carbon capturing technologies

Even if total CO_2 emissions are reduced from the present level of $\sim$700 kg/ton of OPC to 560 kg (see ref. 1 in preface), a reduction of 20%, the emission level will still be higher than the target set, e.g., 525 kg/ton of clinker set by IPCC to be achieved by 2020.

It is therefore necessary to also seek other avenues to achieve this objective. One possibility that is being seriously followed is to capture CO_2 from the exhaust gases and store it safely in liquid form.

1.2.1 Carbon capture and storage (CCS) schemes are being considered for thermal power plants. They envisage separating CO_2 from the flue gases (kiln exhaust gases from cement kilns), collecting it, compressing it, transporting it in liquid form through pipelines and storing it in deep underground reservoirs.

1.2.2 Technologies for compressing and transporting are available. Technologies for capture are under various stages of development. Some have been developed on a small scale but none yet on the scale required for today's cement plants.

1.3 Basically there are three types of carbon capture schemes:

1. pre combustion
2. oxyfuel combustion
3. post combustion.

1.4 Pre-combustion schemes

It is envisaged that fossil fuel is replaced by hydrogen as fuel. However this technology has not yet been tried out in a cement plant. Because hydrogen is explosive it cannot be used by itself. It has to be diluted with nitrogen or steam.

Further, CO_2 from combustion only is eliminated; that from calcination still remains.

1.5 Oxyfuel firing

In this scheme, combustion air is replaced substantially by reasonably pure oxygen. CO_2 rich gas is recycled to moderate flame temperature.

The equipment required includes:

an air separation unit to remove oxygen from the air prior to feeding it to the calciner/kiln;

a system to recirculate CO_2-rich gas back to the calciner/kiln burners; and

a gas treatment plant to dry, clean and compress CO_2 and transport it.

See a flow chart of the scheme in Fig. 3.1.1.

Oxygen can be produced on the plant premises. Cryogenic air separators are well known.

A power intensive air separation unit and flue gas conditioning push up energy and operating costs by 45%.

The scheme also needs a two-stage cooler as in white cement plants so that air does not dilute oxygen for combustion. Sealing at all points (kiln inlet and outlets) is also very important.

1.6 Post-combustion schemes

Two main schemes under development are:
1. amine scrubbing
2. carbonate looping.

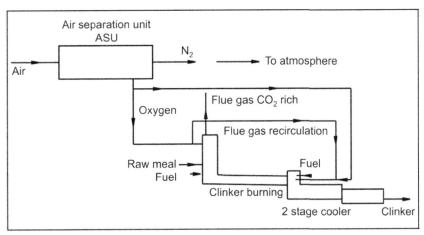

Figure 3.1.1 Carbon capture and storage (CCS) flow chart for oxyfuel firing process. *(Source: PCA R&D Serial No. 3022).*

1.6.1 Amine scrubbing

In this, CO_2 is first absorbed by monoethalomine. It is then separated from the amine solution, dried, compressed and transported to a storage site.

It is easily possible to install the amine capturing equipment in the gas circuit of the cement plant without major modifications. Equipment needed to be installed is shown in flow chart of Fig. 3.1.2.

CO_2 is compressed in a compression plant in which gas is dried and compressed to a pressure of 110 bar and transported in a pipeline. The main disadvantage of amine scrubbing is that it is often degraded by oxygen and impurities like SO_x and NO_x. It is expensive to install.

1.6.2 Carbonate looping:

The basic principle is:

$$CaO + CO_2 = CaCO_3$$
$$CaCO_3 = CaO + CO_2$$

This scheme is in early stages of development. It envisages lime carbonation and calcination. It is based on separation of CO_2 from the combustion gases by using lime as an effective sorbent to form $CaCO_3$.

CO_2 is then separated from the carbonate at high temperatures by using CaO as a regenerable sorbent.

Reverse calcination produces a gas rich in CO_2 and regenerates CaO for subsequent cycles.

Carbonation reaction in the first stage takes place in a fluidized bed combustor at temperatures of $\sim$650–850 °C.

Carbonate particles are separated from flue gas and are sent to the calciner where at $\sim$950 °C CO_2 is separated and taken to storage.

A small amount of fresh sorbent calcium carbonate needs to be added to the system to maintain the overall sorbent activity. In

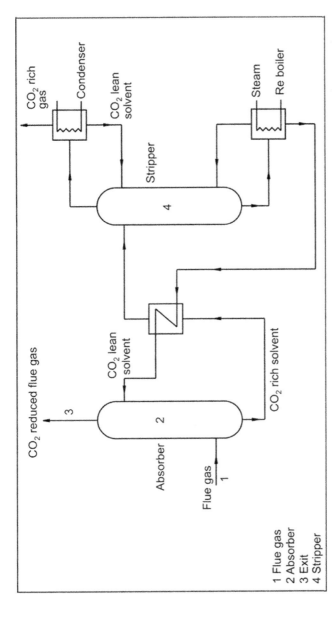

Figure 3.1.2 Carbon capture and storage (CCS) flow chart for amine scrubbing process. *(Source: PCA R&D Serial No. 3022).*

this process of capturing CO_2, it is necessary to deal with mass flows 60 times larger than those required to produce clinker.

Using lime purge materials from the calcium looping process as a raw material substitute in cement production allows considerable savings of fuel and CO_2 emissions. See a flow chart of the scheme in Fig. 3.1.3.

Ironically, more CO_2 emissions occur in systems with CCS because of lower operating efficiency. Further developments are awaited.

1.7 All these processes are used to capture CO_2, which is then compressed and transported in pipelines (over long distances) to underground reservoirs.

Availability of suitable geological rock formation for storing CO_2 is a must for these schemes and adds considerably to the cost of storing CO_2, which could be as high as 50 €/ton of CO_2.

CO_2 is transported in pipeline systems that have moved natural gas and oil for decades. It is injected to a depth of 1-5 km in deep underground rock formations where it is stored permanently.

Geological formations suitable for this purpose include depleted gas or oil reservoirs, saline formations, coal seams and such.

Fig. 3.1.4 shows a flow chart of transporting captured CO_2 for storage in underground reservoirs and in sea beds.

Plate 3.1.1 shows various possibilities for storing captured CO_2.

1.8 There are quite a few projects under various stages of implementation all over the world in the power sector. However it is too early to say whether CCS will be a viable proposal for even large cement plants.

According to Cemex, CCS schemes are not likely to come into general use in cement plants before 2020.

It is unlikely that they would be viable for cement plants of 5000-tpd capacity or less.

1.9 Without infrastructure, operators will not invest in CCS, and without concentrated sources of CO_2 there will be no incentive to develop pipelines and storage facilities.

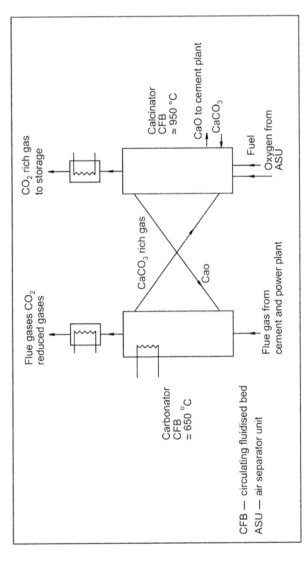

Figure 3.1.3 Carbon capture and storage (CCS) carbonation-calcination loop process. *(Source: Modeling and analysis CO₂ capture in industrial sectors GHG Science & Technology).*

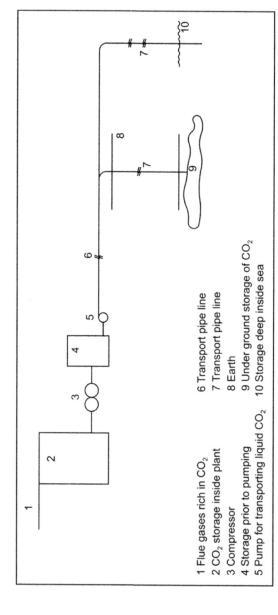

1 Flue gases rich in CO_2 6 Transport pipe line
2 CO_2 storage inside plant   7 Transport pipe line
3 Compressor   8 Earth
4 Storage prior to pumping   9 Under ground storage of CO_2
5 Pump for transporting liquid CO_2 10 Storage deep inside sea

Figure 3.1.4 Flow chart for CO_2 collection, compression, transport, and storage (CCS).

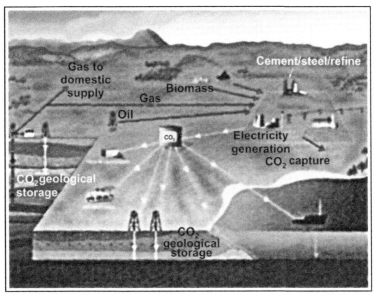

Plate 3.1.1 Schematic diagram of possible CCS systems. *(Source: Technical report TR004/ 2007).*

There is huge energy demand for operating CCS which would have to be taken into account.

Transport and storage has a political angle also.

They have been described here as they could be part of green cement plants in the future.

1.10 Calera process

In this process CO_2 in the flue gases of thermal power plants is led to seawater, where it reacts with the salts in seawater to produce a cement-like product. This has been described in Section 9 on cement substitutes.

This process could possibly be used by cement companies to produce a substitute cement as a byproduct.

RECOMMENDED READING

1 Science Direct CO_2 capturing technologies for Cement Industry By Adina Bosoaga & others

2 CO_2 Capture Cycle Role of CCS in reducing emissions from cement plants in N. America By Mahasenan et al - Pacific Northwest National Laboratories

3 Reduction of GHG emissions by integration of cement plans, power plants and CCS Systems By Luis Romeo et al - Modelling & Analysis- CO_2 capture in Industrial Sector - Greenhouse gases Science and Technology

4 Cement Technology Road map 2009 Carbon emission reduction by 2050 World Business Council for Sustainable Development

5 Carbon Capture and Storage – Cemex's position Cemeex update 2011

6 Carbon Capture Technology ECRA's approach towards CCS European Cement Research Academy By Volker Hoenig & others

7 CCS Technology – Options and Potentials for Cement Industry by Volker Hoenig et al Portland Cement Association- PCA R&D Serial no 3022

8 Carbon dioxide Control Technologies for Cement Industry By Volker Hoenig et al - GCEP Workshop on Carbon Management in Manufacturing Industries- 2001 for ECRA

9 UNIDO Carbon Capture & Storage in Industrial Applications Technology Synthesis Report- Working Paper 2010.

SECTION 4

Alternative Fuels and Raw Materials

Contents

List of Figures

List of Plate

List of Tables

CHAPTER 1

General Introduction

1.1. "Sustainable development and operation" has become the guiding philosophy for all industrial activities, including production of cement.

By this philosophy, development should be eco-friendly and should conserve nonrenewable natural resources to the extent possible.

1.1.1. From both these angles, it is necessary not only for the cement industry but for all industries consuming fossil fuels to start looking for alternatives that will replace fossil fuels progressively.

1.1.2. As discussed in paragraph 1.2.2 of Chapter 2 in Section 1, the carbon in fossil fuels contributes to ~ 0.29 kg of CO_2/kg clinker at the prevailing level of fuel efficiency.

1.1.3. Fuel costs keep increasing year after year, influencing the costs of production of cement. Therefore from this angle also substitutes or alternatives which have commercially usable heat value, which produce a lesser quantum of greenhouse gas (GHG) and which are cheaper are proposals worth considering.

1.2. Types of likely alternative fuels to substitute for coal or oil

The cement industry has been using fuels like petroleum coke and lignite as alternative fuels. Vertical shaft kilns use coke breeze from coke ovens. Oil was extensively used to fire cement kilns. Wherever available gas was also used as fuel.

Technology and equipment to fire all these fuels in kilns is available.

However the present need is to go beyond these and find unconventional sources that satisfy the two criteria mentioned above.

Therefore many industrial wastes and waste products suggest themselves as substitutes, albeit to a limited extent.

Designing Green Cement Plants
http://dx.doi.org/10.1016/B978-0-12-803420-0.00008-1

Some such substitutes have already been in use for quite some time now, such as:

1. used tires inwhole or in shredded form
2. industrial waste oils and lubricants and paints
3. plastic wastes
4. municipal solid wastes (MSW)
5. biomass

Use of wastes serves a dual purpose, first as fuel, and second and equally significant, problems with disposal (the large areas required for landfill) are reduced.

Use of biomass is even more significant as it is carbon neutral. That is, the combustion of these fuels do not add to the GHG emissions.

Thus there are a great many possibilities.

Table 4.1.1 lists desirable properties in possible alternative fuels (AFs).

Table 4.1.1 Desirable Properties in AF

	Parameter	AF—Liquid	AF—Solid
1	Calorific value (kcal/kg)	4000–4400	4000–4400
2	Water (%)	<20	<20
3	Flashpoint (°C)	>60	Not applicable
4	Chlorides (%)	<1.5	<1.5
5	Total halogens (%)	<1.5	<1.5
6	Sulfur (%)	<1.5	<1.5
7	PCBPCT (ppm)	<50	<50
8	Heavy metals (%)	0.2–1.0	0.2–1.0
9	Heavy metals (ppm)		
	TI + CD	<20	<20
	V	<100	<100
	As	<60	<60
	Cr	<400	<400
	Hg	<10	<10
	CD + TI + Hg	<100	<100
	All above + Co + Ni + Se + Te + Sb + Sn + Pb	<2500	<25
10	pH value	5–9	5–9
11	Ash (%)	<5	<25
12	Odor, toxicity	No strong odor, nontoxic	
13	Contamination	Should be free of inerts like grit, rags, glass Stone, metal, etc.	

Table 4.2.1 shows useful heat values of commonly known AFs.

1.3. Problems with using alternative fuels

There are two kinds of problems with beginning to use alternative fuels.

1. There are a great many types of alternative fuels, and also a great many sources for them.

   The properties that influence their use, like calorific value, moisture and composition, vary greatly from source to source for the same type of fuel. For example fuels like MSW and refuse-derived fuel have very different calorific values from time to time from the same source. From different sources there are further variations.

   Availability on a continuous basis from the same source(s) in the quantities required by large cement plants is another important factor.

2. The second problem is related to the combustion of some types of AF. During combustion, they generate exhaust gases that contain undesirable fractions of chlorine, alkalis, dioxins and heavy elements like mercury which affect the continuous operation of the kiln and are harmful to the atmosphere.

   A constant watch must be kept on these aspects while using these fuels.

   A long-term solution would be to install kiln bypass systems in such cases, but this adds substantially to the installation costs of AF systems.

1.4. Properties of AFs

As mentioned above, AFs come from a great many sources and in many different forms. Their properties also vary greatly. It is therefore necessary to find detailed information about them. In Chapter 2 pertinent information collected from different sources is presented.

1.5. Forms of availability of AFs

The forms in which these fuels are received influences design and installation of systems for cement plants to receive, store, extract, meter and fire the fuels in kilns and calciners.

Presently most kiln-calciner systems are designed for firing pulverized coals, oils and gas.

The alternative fuels on the other hand would hardly be available in these forms.

Since AF can replace main fossil fuels partially, it becomes incumbent on the cement plants to prepare two or more types of fuels for firing.

This situation may very well happen not once but again and again, given the diverse locations, sources and types of waste fuels.

Very detailed investigations need to be carried out at the design stage with regard to selection of the type of AF and its sources.

1.6. Selection of AF to be used

The first step to be taken is to establish procedures and tests that will help in selecting the most suitable AF from among a great variety of possibilities.

AFs should be subject to preliminary and comprehensive tests before they are accepted as suitable.

Cement companies will have to carry them out if they are required to pick up waste on an "as is where is" basis.

On the other hand an intermediate service provider may supply AF in ready-to-use form to the cement company. In that case the ISP would carry out these tests.

Figs. 4.5.1–4.5.3 in Chapter 5 furnish in schematic form the various steps and stages in examining waste fuels available and their selection or rejection.

From among the several alternatives available, an AF which has a greater possibility of being consistent in quality and availability would be the obvious choice.

Table 4.2.2 lists possibilities for using alternative fuels. It also shows broad properties such as average useful calorific value, moisture, the form in which they are commonly available, density, etc. It also shows possible sources for the fuels.

A cement plant wanting to use AF has to examine, with regard to its own location and capacity, which types would suit it most.

These aspects are dealt with in Chapter 2.

1.7. Planning to install systems for firing alternative fuels

This will have to begin at the point of receiving the fuel. The system has to be tailor made to suit the selected fuel type.

Broadly, AFs may be received as:
> liquids or sludges
> pellets and/or briquettes
> granular/crushed form.

1.7.1. Storages

First it is necessary to arrive at the daily requirement and the number of days' stock to be maintained.

When received in liquid form, it will have to be stored in tanks, with facilities for preheating, filtering, and pumping.

Received as briquettes or pellets or in granular form, it can be stored in suitably sized sheds or halls with suitable facilities for receipt, like truck unloaders, and for withdrawal, like overhead grab-bucket cranes.

In specific cases the fuels would require fire protection, etc. These aspects are dealt with in Chapter 3.

1.7.2. Processing for use

Most types of AFs require some kind of processing make them ready for use. This is explained in Chapter 4. See Table 4.4.1.

Liquid fuels have to be filtered to remove dirt and foreign bodies and heated to bring their density to a point where they can be fired by a multi-channel burner in either the kiln or the calciner. Metering liquid fuels should not pose a problem.

Specially designed multi-channel burners are available that would simultaneously fire coal, regular fuel, and waste liquid fuel in predetermined proportions.

Pulverized AF can be fired into a kiln by conventional burners.

Calciners are able to receive all kinds of fuels in all forms, liquids, gases, pulverized solids and granular solids.

1.8. Where to fire alternative fuels

Fuel fired should be burnt completely.

In rotary kilns temperatures are high and therefore all kinds of fuels will be completely burnt. Retention times in burning zones are also sufficient. The kiln is thus an ideal incinerator. Ash in fuels would be almost wholly absorbed in clinker. There is thus no problem with its disposal.

Kiln gases enter the precalciner and preheater, and if they contain undesirable compounds like alkalis and chlorides that deposit on the walls of the preheater, continuous operation of the kiln itself would be affected.

In such cases a portion of the gases at the kiln exit would be bypassed and dust collected can be thrown out of the system.

1.9. Viability of using AFs

The viability of using AFs continuously is influenced by
1. uncertainty about quality and consistency
2. uncertainty about continuity of supply
3. possible contamination of waste gases by undesirable compounds and elements and the steps that would have to be taken to contain them.

The use of AFs thus has some inherent constraints:
1. maximum use permissible without adverse impact on quality of cement or environmental pollution
2. mininum use that makes it a commercially viable proposition
3. AF use pushes up specific fuel consumption slightly because of inconsistency in calorific value. It also results in somewhat higher specific gas volume from the preheater.

Above all, use of AFs would be accepted and adopted by the cement industry if it provides financial benefits as well.

1.10. Hazardous wastes

Among the wastes that have combustible value there are also hazardous wastes which are difficult to handle, store and process, as they can catch fire, explode, emit obnoxious gases, or all of these.

The cement industry should proceed cautiously in selecting hazardous wastes as AFs.

1.11. Impact on plant layout

From the foregoing it is seen that the cement plant would have to

1. design and create additional receiving, storing and handling facilities for AFs in, addition to similar facilities for regular fuel like coal
2. select appropriate metering and feeding equipment for AFs
3. select/modify designs of calciners to receive AFs
4. confer with consultants and select and install a bypass system for the kiln if needed
5. introduce additional quality control measures to monitor closely the impact, if any, on the quality of cement clinker and on exhaust gases
6. work out cost and benefits of using AF and monitor them, and change the source and type, or both, of AF used and its proportion accordingly.

Figs. 4.6.1–4.6.9 in Chapter 6 show flow charts for systems using common types of AFs.

Layouts nos. 4.6.10 and 4.6.11 in the same chapter show how the system would be integrated in the plant layout.

1.12. A development that would accelerate the use of AFs would be infrastructure, in the form of intermediate service providers who would set up facilities to collect AFs from their origin, process them and make them available in ready-to-use form to the cement industry.

CHAPTER 2

Properties of Commonly Available AFs

2.1. Having made a case for using Alternate Fuels (AFs) on a priority basis, the next step is to gather as much information about them as possible, with a view to selecting those available that are likely to meet the cement company's (CC) requirements best.

It is also necessary to look into the continuous availability of required quantities of the promising AFs.

2.2. Common types of AFs

Broadly there are four types of AFs.

1. Agricultural wastes and biomasses.
2. Agricultural products cultivated for their value as fuel.
3. Industrial wastes of all kinds that have heat value. They will come from a myriad of industries.
4. Municipal solid wastes (MSWs)—mainly garbage collected from residences and commercial complexes of towns and cities.

   2.2.1 Cognizance should also be taken of **hazardous wastes**, which have heat value but are difficult to handle, store and use as they are inflammable or explosive.

2.3. Properties of common types of AFs

Table 4.2.1 shows representative heat values of commonly available wastes. It can be seen that they are comparable to heat values of fossil fuels, and most have potential as substitute fuels.

Table 4.2.2 shows properties of different AFs in greater detail. It shows which are hazardous and which are carbon neutral.

Broadly speaking, biomass fuels are carbon neutral, in that the CO_2 emitted has already been accounted for.

Designing Green Cement Plants
http://dx.doi.org/10.1016/B978-0-12-803420-0.00009-3

Details like size and moisture, etc. furnished are indicative of further processing required to make them ready for use.

This table shows (indicative only) the extent to which they are likely to substitute fossil fuel.

This table could thus be a good starting point for using AFs.

Tables 4.2.3, 4.2.3a, and 4.2.3b show the extent of possible variations in composition of MSWs in different parts of the world and in different urban complex locations.

Table 4.2.4 shows wastes that are **not** suitable as AFs and why.

Tables 4.2.1–4.2.4 together furnish a starting point and demonstrate the availability of a great many types of wastes with heat value from myriad sources.

2.4. Selection from among the promising wastes

In Table 4.1.1 in Chapter 1, desirable properties, or properties that should be looked for in an AF, were listed.

The table furnished desirable and limiting acceptance values for various characteristics like calorific value and moisture, and also for undesirable elements like mercury, chlorides, etc.

Table 4.2.1 Typical Calorific Values of Different Fuels and AFs

Sr. No.	Fuel	Low Heat Value (kcal/kg)	High Heat Value (kcal/kg)
1	Coal	6600	6900
2	Petroleum coke	7100	7800
3	Waste-derived fuel	5400	6200
4	Waste tires	7500	7900
5	Wood	4700	4900
6	Sawdust	4700	5100
7	Municipal solid waste (MSW)	3200	3600
8	Oil	9600	10,000
9	Waste oil	5000	5200
10	Plastics	9000	
11	Paper	2900	5300
12	Solvents	6000	

Source: Case study manual on AF & R Utilization in Indian Cement Industry by CII.

Table 4.2.2 Properties of Some Common Wastes Used as Alternative Fuels

Category	Name	State	Moisture (%)	Density (kg/m³)	Ash (%)	Carbon (%)	l.h.v. (kcal/kg)	Hazardous/ Nonhazardous	Associated Emissions	Substitution rate up to (%)	Carbon Emission Factor	Net CO₂ Emission
Agricultural biomass	Coffee husk		11		11		3900					
	Rice husk	Solid	5–10	100–150	20	39	3150–3900	Non	Chlorine	35	0.35	Carbon neutral
	Wheat straw	Solid	7–14		5–9	45–49	3800–4300	Non		20	0.42	Carbon neutral
	Bagasse	Sludge	10–15		4	44	3400–4600			20	0.39	Carbon neutral
	Hazelnut shells	Solid	9		3.5	53	4200	Non		20	0.48	Carbon neutral
	Palm nut shells	Solid	10				2850			20	0.36	Carbon neutral
	Wood	Solid	33			50	3700–4100		Chlorine	20	0.36	Carbon neutral
Nonagricultural biomass	Dewatered sewage sludge	Semi solid	75			30–54	2500–6900	Non	Heavy metals	20	0.21–0.39	Negative
	Paper	Solid			8.3	48	3000–5300	Non	Chlorine	20	0.42	Negative
	Sawdust		20		2.6	47	3900	Non	Chlorine	20	0.38	Negative
	Waste wood		33		0.9	50	3700–4100	Non	Chlorine	20	0.33–0.49 0.34	Negative
	Animal waste		15			34	3800–4500	Non			0.29	Negative

Continued

Table 4.2.2 Properties of Some Common Wastes Used as Alternative Fuels—cont'd

Category	Name	State	Moisture (%)	Density (kg/m³)	Ash (%)	Carbon (%)	l.h.v. (kcal/kg)	Hazardous/Nonhazardous	Associated Emissions	Substitution rate up to (%)	Carbon Emission Factor	Net CO_2 Emission
Industrial wastes	Textiles		6			45	3900	Non	Sb, Cr, Zn	30	0.42	
	Plastics						9000					Negative
	Paint residues	Semi solid	9		34	40–50	3900		Dioxin		0.42	
	Spent solvents	Semi solid	10–16			48	5000–6000		Dioxin	<20	0.4	Negative
	Tires	Solid					6700–8900	Non	No_x, SO_2, CO		0.56	
	Hazardous wastes, solid						3500		Dioxin			
	Hazardous wastes, liquid						3750		Dioxin			
Petroleum-based wastes	Polythene polypropylene polystyrene		2.1		27	71	11,000		Chlorine	70	0.7	Negative
	Waste oils		5			46	5200		Zn, Cd, Cu, Pb		0.44	Negative
	Petroleum coke		4			78	4500–8100		SO_2–NO_x	Up to 100	0.78	Negative
	Municipal solid wastes liquid		10–35			40	2900/3800			Up to 30	0.26–0.36	Negative

Source: Case Study Manual on AF & R Utilization in Indian Cement Industry.

Table 4.2.3 Typical Compositions of Municipal Solid Wastes (MSWs)

Sr. No.	Item	Residential Area (%)	Commercial Area (%)	Slum (%)
1	Organic matter	81	69	75
2	Plastics	10.75	12	9
3	Paper	6	8	12
4	Cloth	0.25	5.0	4
5	Metals	0.5	–	–
6	Glass	1.25	–	–
7	E-waste	0.25	–	–
8	Coir	–	1.0	–
9	Coconut shell	–	5.0	–

Source: Co-incineration of MSWs in Cement Industry an article by Mr. Axel Seemann Centre for Sustainable Development in Cement Industry.

Table 4.2.3a Typical Compositions of Municipal Solid Wastes (MSWs) in USA

Sr. No.	Item	%
1	Paper & paperboard	28.5
2	Food scraps	13.9
3	Yard trimmings	13.4
4	Plastics	12.4
5	Metals	9.0
6	Rubber, Leather, Textiles	8.4
7	Wood	6.4
8	Glass	4.6
9	Other	3.4

Source: US Environmental Protection Agency.

Table 4.2.3b Typical Compositions of Municipal Solid Wastes (MSWs) in India

Sr. No.	Item	%
1	Biodegradable matter	48
2	Plastics	9
3	Paper	8
4	Rags	4
5	Glass	1
6	metal	1
7	Inert	25
8	Other	4

Source: National Solid Waste Association of India.

Table 4.2.4 Wastes Not Suitable as AF, processing required to be done

Sr. No.	Waste	Enrichment Pollutant in Clinker	Excessive Emission	Occupational Health & Safety	Potential for Recycling	Negative Impact on Kiln Operation
1	Electronic scraps	x	x		x	
2	Batteries	x	x		x	x
3	Health care wastes					
4	Mineral acids		x	x		x
5	Explosives	x		x		
6	Asbestos wastes			x		
7	Radio active wastes	x		x		
8	Unsegregated municipal solid wastes	x	x		x	x
9	Cyanide wastes		x	x		
10	Fluorescent lamps	x	x			

Source: Case Study Manual on AF & R Utilization in Indian Cement Industry by CII.

Properties of prospective AFs with corresponding values can be compared in this table to make a selection.

2.5. A set of values obtained from another source is furnished below.

Chlorine	<0.3%
Sulfur	<2.5%
Calorific value	Minimum 3500 kcal/kg
Heavy metals	<2500 ppm
Mercury	<10 ppm
Cadmium/thalium	<100 ppm

2.6. Consultants

It may be good idea to engage experienced consultants and seek their advice not only on AF selection but also on the design of a system for processing and using them.

CHAPTER 3

Feasibility of Using Alternative Fuels in Cement Kilns

3.1. Wastes as fuels—savings in fossil fuels

Many types of agricultural and industrial wastes have sufficient heat value.

They can therefore be used as fuel substitutes for fossil fuels which are nonrenewable.

It is in the interests of all concerned to conserve fossil fuels for the benefit of future generations.

Hence wastes have assumed value and efforts are being made by users of fossil fuels to find alternative sources of fuel (AFs).

As far as the cement industry is concerned, at the present level of operational efficiency, 0.16 kg of coal with a useful calorific value of 4500 kcal is required to produce 1 kg of clinker. If it were possible to substitute 20% of coal with alternative fuels, the savings would be 0.032 kg/kg of clinker.

Present production of cement the world over is $\sim$700 million tons per annum.

Assuming a ratio of 1.3 for cement to clinker,

Clinker produced per annum $= \sim 540$ million tons.

Coal saved per year $= \sim 17$ million tons.

This is a savings that cannot be ignored.

3.2. Savings in greenhouse gas emissions

Burning of coal contributes $\sim$0.29 kg/kg clinker produced.

If coal is replaced by AF, GHG emissions would also be reduced because of the lower percentage of carbon in waste fuels. Some types

Designing Green Cement Plants
http://dx.doi.org/10.1016/B978-0-12-803420-0.00010-X
Copyright © 2016 BSP Books Pvt Ltd.
All rights reserved.
Published by Elsevier Inc.

of waste fuels, like biomass, are carbon neutral, that is, CO_2 generated in their combustion has already been accounted for.

Let us say that 0.032 kg of replacement AF has 20% biomass. The balance of 80% has 40% carbon. The CO_2 generated would be:

$$0.032 \times 0.8 \times 0.4 \times 3.66 + 0.032 \times 0.2 \times 0 = 0.037 \text{ kg/kg clinker}$$

Total CO_2 generated would be 0.271 kg/kg clinker, a savings of ~6.5%.

Thus if CO_2 released per ton of cement is 615 kg/ton, by using AF it will be reduced to 600 kg/ton of cement, a reduction of ~2% or **14 million tons per annum.**

Reduction of GHG emissions would be higher for a higher percentage of biomass in the AF.

The two kinds of savings make a strong case for the use of AFs. Further, AFs would be cheaper than fossil fuels.

3.3. From Table 4.2.2 it can be seen that the majority of non-biomass wastes have carbon emission factors of less than 1. Hence it is desirable to substitute fossil fuels with alternative fuels to the extent possible.

3.4. Can wastes be used as fuels in cement kilns?

1. Commonly wastes are either dumped into outlets like rivers or the sea or are dumped as garbage in municipal areas. Municipalities have to collect them and dispose of the garbage in landfills far from the residential, commercial, and industrial areas. However as towns grow dumps come too close to the expanded towns.

2. Nowadays municipalities install incinerators to burn off wastes (MSW) in most cases; the heat generated is not used and exhaust gases often cause pollution. The best way is therefore to burn them off in a controlled manner whereby the heat generated can be put to effective uses like heating water, generating power, and producing cement clinker when burnt in kilns.

3. Fortunately the temperature profile of cement kilns and the residence time therein are such that almost all types of fuels and waste fuels in all forms, solid, liquid, or gas, can be burnt in them.

In kilns with calciners, fuels can also be burnt in calciners,

4. Liquid and gaseous fuels can be easily burnt in kilns and calciners using multi-channel burners. Pulverized solid fuels can be burnt similarly. Solids, coarsely ground or as briquettes or pellets, can be fired in calciners. Solids in shredded form, like shredded tires and plastics, can be similarly introduced into calciners.

Equipment like the "Hot Disc" for feeding whole tires at the kiln inlet end and "Gunnax" for injecting them into the kiln from the kiln hood is now available. Shredded tires can be introduced at the kiln inlet end or in calciners. Sludges can be briquetted and fired in calciners.

5. Ashes from AF get mixed with clinker in the same way as ashes from coal. There is no adverse effect on the quality of clinker produced.

3.5. Possible difficulties with firing wastes and how they can be overcome

1. AFs are generally fired up to 30, maximum 40%. Thus, AFs are fired along with regular and conventional fuels.

A certain amount of close monitoring is required to maintain temperature profiles because the AFs are not as consistent in quality (calorific value, density, moisture, etc.) as coal (and of course oil) that has been processed for firing in the cement plant itself.

2. While using some AFs close watch must be kept on exhaust gases for excessive alkalis, chlorides, and NO_x that can cause difficulties in operation.

Care is also required to monitor heavy metals and dioxins. In extreme cases it may become necessary to bypass a small amount of gases from the kiln inlet end.

3. To keep down NO_x calciners may have to be designed or modified to provide for multistage firing of fuel and multistage introduction of raw meal.

4. Retention time in calciners can be increased for completing combustion by introducing fluid beds.

5. Secondary calcining can also be done.

All the above-mentioned new designs are available and have been in use at many places for considerable time. Hence there should be no problem using them in AF systems.

3.6. Therefore, the feasibility of using different types of AFs and the small modifications or new designs required for pyroprocessing systems are well established.

Summing up, feasibility of using various types of AFs has already been proved.

European countries have reached a stage of **zero fuel costs** by using AFs. Even a less developed a country like Ethiopia is beginning to use biomass to substitute for fossil fuels in its cement kilns.

3.7. Rates of substitution

Very high rates of substitution, 40% and more, can be achieved if a tailored pretreatment and surveillance system is in place.

CHAPTER 4

Possibilities of Using AFs in Cement Plants

4.1 Having established the necessity (desirability) and feasibility of using Alternate Fuels (AFs), the next step is to explore actual possibilities for using these alternative fuels to substitute for conventional fossil fuels in cement kilns and calciners.

The actual possibilities depend on many external factors over which cement companies (CC) have little or no control.

4.2 Availability

The most important factor is the continuous availability of a given AF in sufficient quantity.

Here the canvas is not just one cement plant but all the cement plants in a specific area.

For example, an area may have a number of cement plants. It may also have rice mills producing rice husk which has been found to be suitable as a secondary fuel.

The rice husk (a seasonal commodity) should be available in sufficient quantity to substitute for 20% of the coal used by cement plants in the region. Its availability year round should be ascertained in detail.

4.3 Preprocessing

A great variety of alternative fuels are wastes of agriculture, industry, and townships.

While they have heat value high enough for combustion in cement plants, the forms in which they are available from respective sources are unlikely to be readily usable for firing in cement kilns/calciners.

4.3.1 Some may be available in irregular lumps/broken or crushed form/powder or all of these together. It is difficult to fire such material while maintaining the desired temperature profile in

Designing Green Cement Plants
http://dx.doi.org/10.1016/B978-0-12-803420-0.00011-1

the kiln and calciner. Others may be available in liquid form or as sludges with varying degrees of consistency.

Besides uniformity in heat value, there should be some uniformity in density, size, and moisture.

Table 4.4.1 shows the preprocessing required for various types of AFs to make them ready to use.

4.3.2 The plants producing these wastes are not interested in processing them to make them suitable for cement kilns or for anybody else.

An intermediate agency is thus often required to collect waste, process it and deliver it to cement plants and other industries using it.

4.3.3 Intermediate Service Providers (ISP)

If the demand is sufficient, entrepreneurs come forward to provide the services of collecting waste, processing it and delivering it to cement plants in the region.

Common forms/types of preprocessing required would be:

- Drying
- Crushing/shredding
- Briquetting

4.4 Procurement

The next step is to bring the wastes to the doorsteps of cement plants on a regular basis in the quantities required, so that cement plants do not have to maintain large stocks of alternative/secondary fuels in the plant in addition to the stocks of regular fuel.

In short, the actual possibility of being able to use AF and the extent to which it can be used very much depends on the infrastructure available in the region.

4.4.1 All types of waste fuels are not found in all places. As mentioned above, a region may contain both cement plants and rice mills. The ISPs collects, treats and delivers rice husk as secondary fuel to cement plants in the region.

Table 4.4.1 Processing Required to Prepare Alternative Fuels for Firing

Sr No.	Category	Description	Commonly Found As	Cleaning Filtering	Preheat	Drying	Crushing	Grinding	Shredding	Briquetting Pelleting	Gasification	Final Form
1	Liquids	Liquids	Effluents, waste oils	x	x							Liquid
		Sludges	Industrial Sludges & waste paints	x		x				x		Briquettes, Pellets
2	Solids	Irregular lumps Dry		x			x		x			Sized, Shreds, Powder
		Irregular lumps Wet		x		x	x	x	x			Sized, Shreds, Powder
		Granular to Fine—dry		x			x	x	x			Shreds, Powder
		Granular to Fine—wet		x		x		x	x	x		Shreds, Powder, Briquettes
		Whole pieces	Tires						x			Shreds, Whole
3	Biomass	Some forms of biomass		x		x		x			x	Gas

x Processing required

4.4.2 Large cities produce huge quantities of wastes (municipal solid wastes, MSWs). They also need to be treated to make them ready for use.

Each type of waste has to be treated differently.

Cement plants prefer to use one particular type of alternative fuel (or maybe two) as they cannot install handling and feeding equipment that suits a number of different types of AFs.

4.4.3 Cement plants have to assess the situation in terms of the alternative fuels available in the vicinity and select from among them those that ensure uninterrupted production of quality cement without adverse impacts on the environment.

4.5 Costs of AFs

While waste fuel itself may be free, cement plants have to pay the costs of processing and transport.

Transport costs may be significant depending on the distance the AF is brought.

Biomass/agricultural fuels may be available within a short distance for many cement plants. MSWs and industrial wastes may have to be transported over 200 km to bring them to cement plants.

4.6 Hazardous wastes

Hazardous wastes can also be used as fuel. Special precautions are needed in handling and storing them. Only firms specialized in handling hazardous dusts should be engaged in procuring them as fuel.

4.7 While CC may not like to take on the responsibility of preparation and procurement of AF, they could come together to promote ISPs.

CHAPTER 5

Procurement, Processing, Storage, and Transport of AFs

5.1 Most wastes have to be collected from their origin. Sometimes requirements cannot be met from one source. It is then necessary to collect them from different locations spread over a wide area.

5.2 Rigorous testing and selection criteria

Whether procurement and processing is done by a cement company or by an intermediate service provider (ISP), AFs proposed for use should be subjected to a rigorous testing and selection/rejection procedure.

5.2.1 Some tests have to be done at the source only. Others are done on the premises of the ISP or CC (cement company).

Tests should also evaluate the preprocessing necessary to make the AF suitable for use.

5.2.2 Further, because AFs come from different sources and, being wastes, are inherently inconsistent in quality, the testing procedure has to be continuous so that the AF selected does not have to be rejected at the point of use.

5.2.3 Flow charts in Figs. 4.5.1–4.5.3 lay out procedures and steps for testing AFs at the source and in laboratories of the ISP and CC. If followed, rejection and hence nonavailability at the point of use will be avoided.

5.3 Preprocessing

In most cases it is necessary to clean AFs by removing foreign bodies like stone, rags and the like, and obnoxious materials. This is best done at the source.

5.3.1 Most wastes need to be prepared for use as waste fuels in cement kilns. The kind of preparation depends on the waste itself.

Designing Green Cement Plants
http://dx.doi.org/10.1016/B978-0-12-803420-0.00012-3

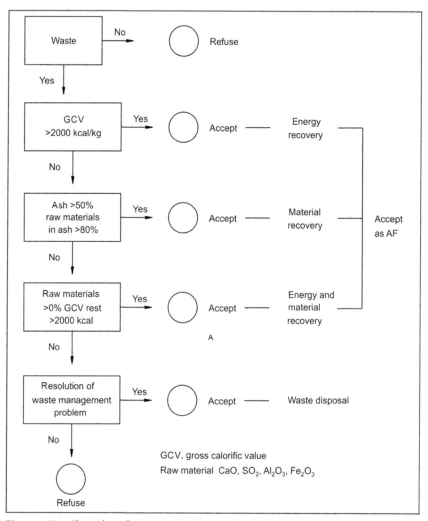

Figure 4.5.1 Flow chart for acceptance/rejection of waste proposed for use. *(Source: CII—Case study manual on AFR utilization in Indian cement industry).*

Preprocessing required for different types of AFs is shown in Table 4.4.1 in Chapter 4.

Preparation involves sizing, drying, mixing, and making into pellets, briquettes, etc.

5.3.2 Relatively dry biomass may be crushed and ground and fired with pulverized fuel. A better way, often adopted, is to make "producer gas" by gasifying the biomass, which can then be easily fired.

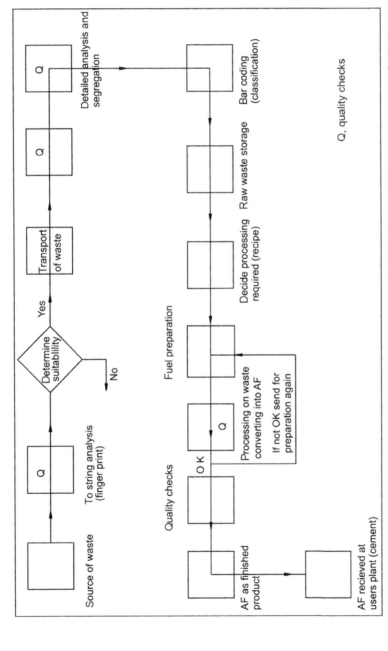

Figure 4.5.2 Steps to prepare waste for AF use by intermediate service provider (ISP). *(Source: GEPIL waste preparation facility).*

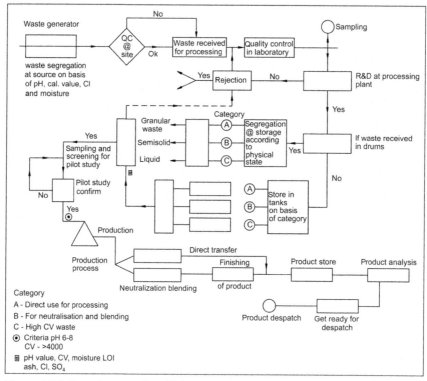

Figure 4.5.3 Steps in preparing AF from waste in greater detail. *(Source: GEPIL waste preparation facility).*

A gasifier can be installed alongside the biomass raw storage. New plants have to install gasification plants and gas burners.

5.3.3 Biomass prepared in the form of pellets and briquettes can be transported by belt conveyors.

5.4 Storage

A very important aspect of using AFs is their storage. Problems connected with storing and handling of oil and coal have been sorted out as they have been the principal fuels for cement and many other industries.

AFs are comparatively unknown and are of such a wide variety that storage systems have to be tailor made.

5.4.1

1. AFs in liquid form may be transported in drums and stored in covered sheds, or alternatively received in tankers and

stored in large tanks like regular fuel oil. Tanks may have to be provided with heaters for extraction and subsequent firing, as in the case of typical oil-fired kilns.

2. AFs in form of sludge may be transported in tipping trucks. Sludge in most cases is dried in rotary driers with paddles/chains inside for movement inside the dryer. Dried sludge will mostly be in powder form. It can be stored in silos and conveyed mechanically/pneumatically inside the plant.

   If the sludge is coarse, it is made into briquettes by passing it through presses. The briquettes are sun dried before use.

3. AFs in solid form need covered storage and handling facilities like additives in the cement plant.

5.4.2 Special attention needs to be paid to designing storage facilities for AFs, in particular biomasses, because of their properties. Biomass AFs have low bulk density and need more space for same tonnage. They tend to absorb moisture and deteriorate in quality with time while in storage.

5.4.3 Hazardous fuels

Some waste fuels are hazardous. They may catch fire easily and or explode or emit obnoxious gases. Their handling may be dangerous as they may cause burns.

Storage for hazardous materials have to be specially designed taking into account their specific characteristics. Provisions for preventing and quenching fires have to be made. Personnel handling fuel have to use protective gloves and wear gas masks.

5.4.4 The best storage option is for the AF to be kept inside the cement plant for the minimum time, just stocked for a day or two. In this case, however, delivery would have to be regular, punctual, and in the quantities required.

Even this small quantity is often stored outside the main plant.

5.4.5 The intermediary company (ISP) that collects and prepares the AF takes responsibility for maintaining stocks of AF on its premises. Users (cement companies) have to assure the intermediary company regular and steady demand.

Table 4.5.1 Various Stages in Receiving, Storing, Processing, and Storing Alternate Fuels

| Category | Description | Commonly Found as | Before Processing | | Equipment for Processing | After Processing | | | Feeding Arrangements | | |
			Transport	Storage		Transport	Storage	Storage	Metering Device	Fed to
Liquids	Liquids	Effluents, waste oils	Tankers	Tanks	Filters, pumps, heaters	Pipeline	Tank	Day tank	Burner	Kiln, calciner
	Sludges	Industrial sludges, waste paints	Tipping, trucks	Covered shed	Dryer, briquetting press	Belt elevator	Covered shed	Day bin	Belt feeder	Calciner
Solids	Irregular lumps, dry		Trucks, wagons	Covered shed	Crusher, shredder	Belt elevator	Covered shed	Day bin	Belt feeder, Hot Disc	Calciner
	Irregular lumps, wet		Trucks, wagons	Covered shed	Dryer, crusher, shredder	Belt elevator	Covered shed	Day bin	Belt feeder, Hot Disc	Calciner
	Granular to fine—dry		Trucks, wagons	Covered, shed	Grinder, pulverizer	Belt elevator	Silo	Day bin	Rotary feeder, burner	Kiln, calciner
	Fine—wet		Wagons	Shed	Grinder, pulverizer	Elevator			Feeder, burner	Calciner
	Whole pieces	Tires	Trucks, wagons	Open	Cleaning, shredder	Crane/chain conveyor belt	Covered shed	Bin when shredded	Gunnax for whole	Kiln hood
									Hot Disc belt feeder when shredded	Calciner
Biomass Hazardous wastes			Trucks, wagons	Covered shed [a]	Gasifier	Pipeline	Tank Covered shed [a]	Tank	Burner	Kiln, calciner

[a] Shed to be equipped with fire extinguishers.

5.5 Testing and quality control

As mentioned in paragraph 5.2 above, it is necessary to test samples of AF to ensure quality before taking delivery and before processing it. This is a continuous process. The expenditure on testing may be sizable as compared to that for common coal or oil.

5.6 Table 4.5.1 summarizes the various stages of preparing various types of fuels for use.

5.7 Firing AFs in the kiln and/or calciner

1. AF in solids irregular in form and size may be crushed or shredded to bring them within a small range of size for uniform burning. They may be stored in covered sheds and conveyed by mechanical conveyors.
2. AF received in granular form like rice husk may be either briquetted or fired as is.
3. Solid AFs may also be ground with coal in coal mills. The proportion is decided taking into account AF composition and its calorific value. The pulverized mix can be fired in a kiln and/or calciner.
4. Tires are often used as waste fuels. In large kilns, whole tires are fired at the kiln discharge end by means of the specially developed Gunnax system of FLS. Otherwise they are fed at the inlet end of the kiln by Hot Disc, also developed by FLS. Tires can also be shredded and fired at the kiln inlet and/or in the calciner.
5. Coarsely ground/granular AF can be fired as received in calciners with longer retention times and fluid beds.

CHAPTER 6

Design and Engineering of Systems for Firing Alternate Fuels

6.1 Consultants

A cement company (CC) may appoint a consultant to help choose the alternate fuels (AF(s)) and also to design and engineer the AF system after the choice has been made.

6.1.1 Among the factors to be looked into closely by the consultant would be:

1. Quality of the AF: its useful heat value and its chemical composition with special reference to chlorides, heavy metals, and other elements/compounds that may adversely affect the quality of cement and or generate objectionable emissions.
2. Consistency in characteristics like heat value, density, moisture.
3. Hazards or risks in transport, storage, and use.
4. Availability of sufficient quantity when required.
5. Kind of pretreatment and processing required to make suitable for firing in kiln/calciner.
6. Cost of AF as received and processed, compared to cost of main fuels used on cost/kcalorie basis.

6.1.2 Consultant should advise the entrepreneur of the impact of using particular AFs on:

Extent to which AF can replace main fuel without adverse impact on quality of clinker or environment

Sp. fuel consumption

System profile of gas volumes, temperatures, and pressure drops

Emissions in exhaust gases

Reduction in GHG emissions

Fuel costs on as-fired basis

Designing Green Cement Plants
http://dx.doi.org/10.1016/B978-0-12-803420-0.00013-5

Capital costs for installing facilities for storage, retrieval, and feeding AF alongside main fuel

6.2 It is obvious that cement plants located in different countries or different parts of the same country will have different options to select from for the AF best suited to them. The exercise outlined above, whether carried out by the entrepreneur or by a consultant, will help with making the correct choice.

6.3 In some locations it may be feasible to use two AFs. The plant may decide to install facilities for both types and use them simultaneously or use one as a standby. However, most plants would opt for one type of AF and equip themselves to use it. This stands to reason also from the point of view of operation and quality control.

6.4 AF systems

The most preferred option would be to get AF in ready-to-use form at the plant site.

The AF system in the plant would then be confined to:
1. receiving AF
2. storing it in transition stage
3. retrieval and transport to kiln/calciner
4. system for storing in day bins—extraction and metering by suitable feeders (adjustable rates of feed)
5. monitoring feed rate along with feed of conventional fuel

6.4.1 If a plant has to use more than one AF, there would be as many storage and retrieval systems, common conveying systems, and as many storage and feeding systems.

6.5 The plant itself can work out system flow charts and system specifications or get a consultant to do this.

Consultants can also help by sizing components like conveyors, feeders, and dust-collecting equipment, taking into account the characteristics of the AF.

6.6 When pre-processing by the cement plant is required, it will receive AF from the transporter and take responsibility for all the stages thereafter.

6.6.1 In such a case, it would be best to store raw AF outside the plant, clean it, and filter it in an adjoining area and pass it through preparation stages like drying, crushing, and briquetting, and transport only processed, ready-to-use AF into the plant.

6.6.2 Hazardous wastes need to be given special attention at every stage, beginning at transport, handling, and storing, then processing and finally feeding it. Precautions are required to deal with possible explosions and fires and harmful gases, such as fire extinguishers and dousers. Workers coming into contact must wear protective gloves, shoes, and masks.

6.6.3 Special attention needs to be given to quantities to be stored and facilities provided in storage areas to protect the material from wind, rain, etc.

6.7 System flow charts and layouts

Flow charts are the best way to depict the design of a system. Hence a flow chart corresponding to the system needs to be drawn as a first step.

Alongside flow charts it is necessary to make layouts of the AF system, showing location and areas required and their integration into the main plant.

6.7.1 Typical flow charts and layouts

Typical flow charts and layouts with explanatory notes are attached. See Figs. 4.6.1–4.6.11

Since there is a large variety to types of AF and how they occur, it is difficult to cover every category.

6.8 Specifications

The next step is to draw out specifications for required machinery.

1. First, properties of materials have to be spelled out, such as:
   particle size (maximum, % fines, etc.)
   moisture
   bulk density
   capacity in tph
2. For feeders, range of variation in feed rate and speeds
   numbers
3. Type of conveyor and feeder most suited

6.8.1 Annexure 1 shows the procedure for working out capacities of the components of an AF firing system.

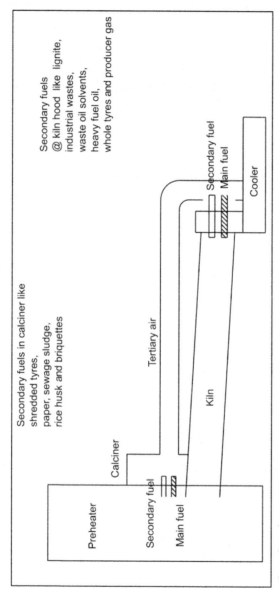

Figure 4.6.1 Flow chart for common AF locations in a kiln-preheater-calciner system.

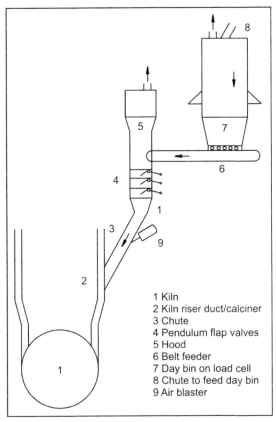

1 Kiln
2 Kiln riser duct/calciner
3 Chute
4 Pendulum flap valves
5 Hood
6 Belt feeder
7 Day bin on load cell
8 Chute to feed day bin
9 Air blaster

Figure 4.6.2 Flow chart for feeding shredded AF into calciner/kiln riser duct.

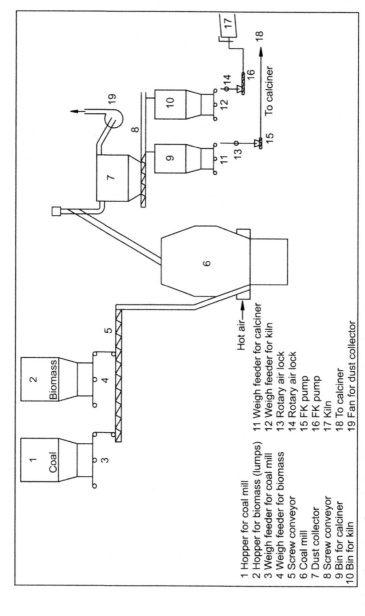

Figure 4.6.3 Flow chart for grinding biomass as AF with coal and firing it in kiln and calciner.

1 Hopper for coal mill
2 Hopper for biomass (lumps)
3 Weigh feeder for coal mill
4 Weigh feeder for biomass
5 Screw conveyor
6 Coal mill
7 Dust collector
8 Screw conveyor
9 Bin for calciner
10 Bin for kiln
11 Weigh feeder for calciner
12 Weigh feeder for kiln
13 Rotary air lock
14 Rotary air lock
15 FK pump
16 FK pump
17 Kiln
18 To calciner
19 Fan for dust collector

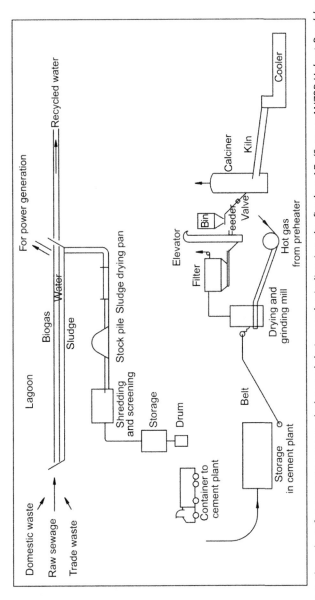

Figure 4.6.4 Flow chart for preparing waste sludge and drying and grinding it to be fired as AF. (*Source: ANZBP Hobart Roadshow 2011*).

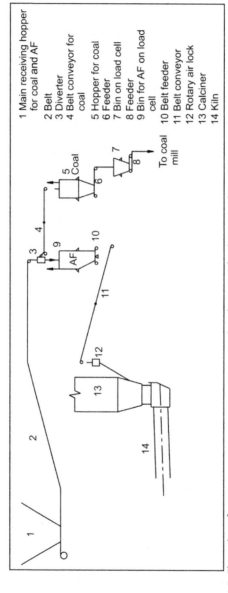

Figure 4.6.5 Flow chart for a common system to receive coal and AF and feed AF to calciner.

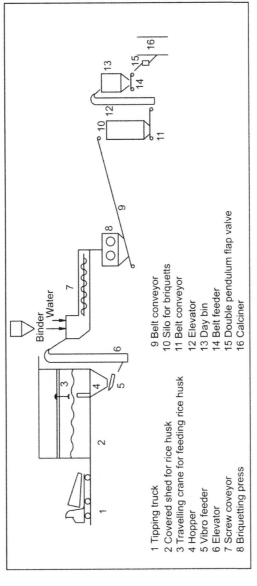

1 Tipping truck
2 Covered shed for rice husk
3 Travelling crane for feeding rice husk
4 Hopper
5 Vibro feeder
6 Elevator
7 Screw coveyor
8 Briquetting press

9 Belt conveyor
10 Silo for briquetts
11 Belt conveyor
12 Elevator
13 Day bin
14 Belt feeder
15 Double pendulum flap valve
16 Calciner

Figure 4.6.6 Flow chart for rice husk as AF, made into briquettes before firing.

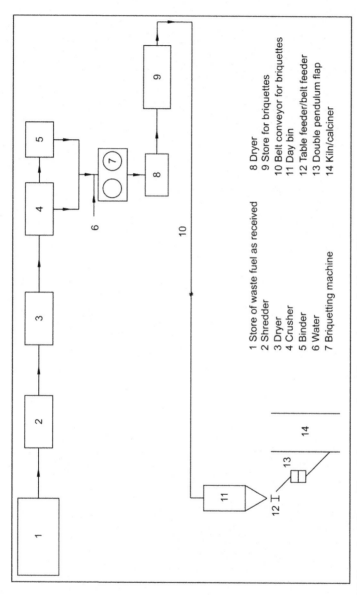

1 Store of waste fuel as received
2 Shredder
3 Dryer
4 Crusher
5 Binder
6 Water
7 Briquetting machine

8 Dryer
9 Store for briquettes
10 Belt conveyor for briquettes
11 Day bin
12 Table feeder/belt feeder
13 Double pendulum flap
14 Kiln/calciner

Figure 4.6.7 Flow chart for making briquettes of an AF to make it suitable for firing.

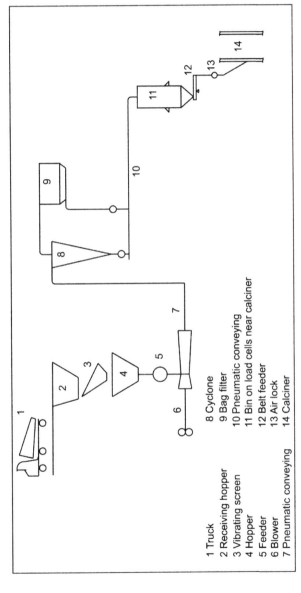

1 Truck
2 Receiving hopper
3 Vibrating screen
4 Hopper
5 Feeder
6 Blower
7 Pneumatic conveying

8 Cyclone
9 Bag filter
10 Pneumatic conveying
11 Bin on load cells near calciner
12 Belt feeder
13 Air lock
14 Calciner

Figure 4.6.8 Flow chart for firing rice husk as received in precalciner.

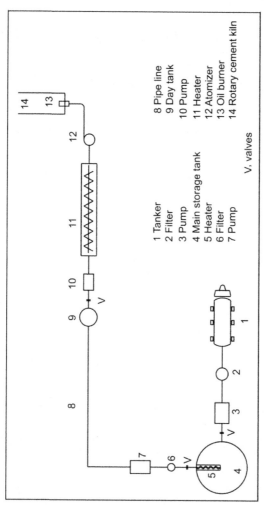

Figure 4.6.9 Flow chart for firing waste oil as AF in a cement kiln.

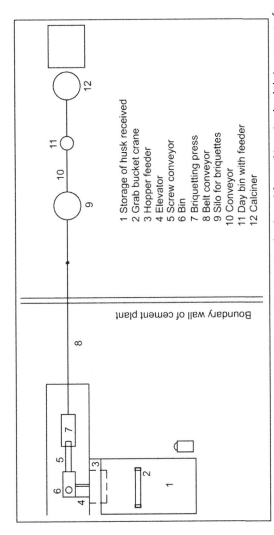

1 Storage of husk received
2 Grab bucket crane
3 Hopper feeder
4 Elevator
5 Screw conveyor
6 Bin
7 Briquetting press
8 Belt conveyor
9 Silo for briquettes
10 Conveyor
11 Day bin with feeder
12 Calciner

Boundary wall of cement plant

Figure 4.6.10 Layout showing arrangement for receiving and storing rice husk and for making rice husk briquettes to feed into calciner.

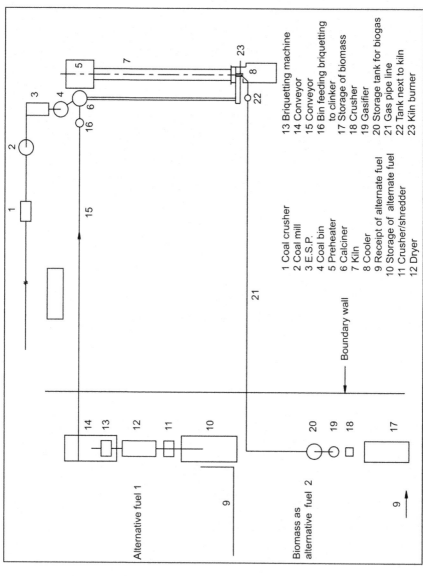

1 Coal crusher
2 Coal mill
3 E.S.P.
4 Coal bin
5 Preheater
6 Calciner
7 Kiln
8 Cooler
9 Receipt of alternate fuel
10 Storage of alternate fuel
11 Crusher/shredder
12 Dryer

13 Briquetting machine
14 Conveyor
15 Conveyor
16 Bin feeding briquetting
to clinker
17 Storage of biomass
18 Crusher
19 Gasifier
20 Storage tank for biogas
21 Gas pipe line
22 Tank next to kiln
23 Kiln burner

Alternative fuel 1

Biomass as
alternative fuel 2

Boundary wall

Figure 4.6.11 Layouts showing arrangements for firing briquettes in calciner, gasifying biomass as AF, and firing it into kiln.

ANNEXURE 1

Workout for Arriving at Capacities for a AF Firing System. AF—Rice Husk

Item	Unit	Entity
Clinkering capacity	tpd	10,000
Design margin	%	10
Sp. fuel consumption	kcal/kg	700
Calorific value coal	kcal/kg	4000
Fuel fired in kiln	%	40
Fuel fired in calciner	%	60
Working hours	no	24
Clinkering capacity per hour with design margin	tph	458
Coal in kg/kg clinker	kg/kg	0.175
Total coal per hour	tph	80.21
Coal in kiln	tph	32.1
Coal in calciner	tph	48.1
AF to be used		Rice husk
To be fired in		Calciner
Useful calorific value	kcal/kg	3500
Moisture	%	10
Bulk density	kg/m^3	150
AF @ 20% of total	%	20
Margin	%	5
Heat supplied	kcal/h	320,833,333
Heat supplied to kiln	kcal/h	128,333,333
Heat supplied to calciner	kcal/h	192,500,000
When AF fired up to targeted capacity heat to be supplied in calciner 20%	kcal/h	64,166,667
With 5% margin	kcal/h	67,375,000
Rice husk to be fired	kg/h	19,250
	tph	19.25
	say	20
Rice husk to be treated and processed		
Rice husk to be made into briquettes		
Wet rice husk with 10% moisture	tph	22
Daily requirement of wet rice husk	t	528
Stock of wet husk		
No. of days	no	3
Stock to be kept	t	1584
	say	1600

Continued

Item	Unit	Entity
Volume of stock pile	m^3	10,667
Dimensions of		~11,000
Triangular pile for		
40° angle of repose	Width × length	
	m × m	20 × 140 (see Table 7.5.3)
Rice husk received in	h	8
Handling capacity	tph	66
Screen capacity	tph	79.2
	say	80
Briquetting press	tph	80
Belt conveyor	tph	100
Hopper/bin	tons	250
Belt feeder	tph	25
0-25 tph		
Double pendulum flap		1

Reference flow chart Fig. 4.6.6.

CHAPTER 7

New Machinery to be Installed for AF Systems

7.1 This may be divided into three broad sections

(1) machinery required to bring AFs to the plant

(2) machinery required for processing AFs into ready-to-use form

(3) Xmachinery for storage, extraction, conveying and feeding AFs into kiln and calciner

The CC has to arrange for all three of the above to convert AF into ready-to-use form.

However if an ISP is delivering the AF, the CC has to arrange for item 3 only.

7.2 Types of machinery

Broadly, machinery can be divided into:

(1) Machinery required to bring AFs from their origins to the plant.

It would mainly consist of tankers, self-tipping trucks and the like. At the plant truck unloaders may be needed.

(2) Machinery required in storage facilities.

This consists of overhead traveling cranes. Walking floors are a new concept.

(3) Machinery required for processing, including machinery for intermediate handling

It would consist of:
- size reduction and separation: crushers, shredders and screens
- reduction of moisture: dryers, rotary/fluid bed
- briquetting: briquetting presses
- grinding: small mill. This can also be done in the main mill.
- filters for liquid AFs
- gasifier (biomass is often best converted into gas before use as AF).

This is described in Table 4.4.1 in Chapter 4.

Designing Green Cement Plants
http://dx.doi.org/10.1016/B978-0-12-803420-0.00014-7
111

(4) Machinery for metering and firing

This is governed by the final form of the AF after processing and includes:

- weigh feeders, solid flow meters, belt feeders, rotary vane feeders and the like
- Hot Disc and Gunnax for whole tires
- feeders with variable speed drives to regulate rate of feed.
- air locks of suitable type for the feeding system
- multi-channel burners that fire both the main fuel and the AF, whether pulverized, liquid or gas

This is presented in Table 4.5.1 in Chapter 5

(5) Machinery for handling and conveying
- This would consist mainly of belt conveyors, screw conveyors and bucket elevators.

7.2.1 When AFs are processed by the CC, there is double handling.

First, when material is received and stored raw and then fed to processing machinery, and

secondly when it is conveyed after processing to final small storage within the plant for firing. (Small storage may be provided after processing to serve as buffer stock and maintain continuity of supply when processing equipment is not working.)

This part of the system is installed outside the main plant.

7.2.2 Suitable dust collecting equipment needs to be installed at appropriate points.

Fire extinguishing equipment is installed when dealing with hazardous wastes.

7.3 Feeding and firing of AF

AF can be fired in the kiln or in the calciner.

Since AF replaces the main fuel only partially, equipment installed for AF works along with that for main fuel and is therefore fitted into the layout so that both can be monitored easily.

7.3.1 The most common points for introducing AF into the kiln-pre-heater-calciner system are as follows.

In the kiln

(1) At the kiln discharge end along with the main fuel, through the same burner

(when the AF is in liquid, gas and or pulverized form). A separate burner may also be used.

(2) Through a system of double gates and Hot Disc. This is suitable for firing whole tires at the kiln inlet end.

(3) At the kiln hood through a Gunnax system

In the calciner

(1) If in granular or shredded form chuted down into the calciner. Even pulverized fuel can be chuted down.

Belt weigh feeders are used to meter the quantity to be fired.

(2) (In calciners designed to limit NO_x) split and introduced in the calciner at two levels (in some designs even raw meal is introduced at two levels).

Suitable airlocks in the form of either pendulum flaps or rotary vanes are necessary, as for the main fuel.

7.3.2 Thus, mostly equipment commonly used for main fuels is required in an AF firing system. It has to be selected and may be slightly modified to suit the form of AF used.

The major equipment specific to the AF includes:
- Hot Disc and Gunnax
- multi-channel/multi-fuel burners

7.4 With specific regard to AF, seldom will any plant would begin to fire the maximum possible quantity of AF immediately. Most likely, quantity will be increased progressively after stabilizing the process at each level.

However, it is prudent to install equipment corresponding to the final targeted capacity.

Capacities are smaller, as the AF fired normally does not exceed 40% of the main fuel.

7.4.1 It can also happen that more than one type of AF is fired simultaneously. Design and planning of such a system needs to be done carefully.

7.5 See Plate 4.7.1 for briquetting press below. In Chapter 7 of Section 7, see:

Plate 7.7.8 for a multi-fuel burner
Plate 7.7.9 for Hot Disc
Plate 7.7.10 for the Gunnax system for whole tires
Plate 7.7.11 for the Gunnax canister system

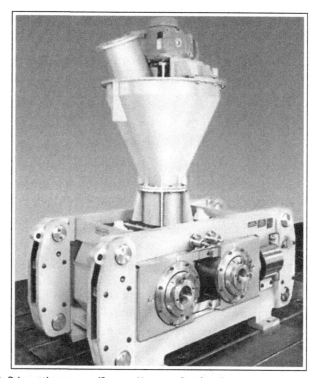

Plate 4.7.1 Briquetting press. *(Source: Koppern Brochure).*

CHAPTER 8

Capital Costs, Payback Period and Benefits

8.1 Capital costs of installing facilities for AF use

Capital costs incurred installing facilities for AF use depend on:

(1) Characteristics of the AF available and proposed for use. This is because some AFs require more processing and preparation than others.

(2) If it is possible to get on site AF in ready-to-use form, then the cement company (CC) does not have to invest in all the facilities for receiving, storing, processing and getting it in ready-to-use form. There will be a higher price, however, for processing done by the intermediate service provider (ISP).

On the other hand, if the cement plant has to pick up the AF on an "as is where is" basis, it will incur capital costs for procurement, storage and processing in preparation for use.

(3) Facilities and equipment to be installed at the plant related to storage, transport and feeding systems designed. These will differ for different AFs.

For example, feeding whole tires requires heavy investment when compared to feeding dry, pulverized fuel or using liquid and gaseous fuels.

(4) While most of the feeding equipment may be available locally, feeders like Hot Disc and Gunnax can only be obtained from specific manufacturers.

(5) If using available AFs also requires installing a bypass system complete with its own cooling and dust collection systems, capital costs increase sharply.

(6) It is also necessary to install quality check systems both in the plant and at the source to ensure that AFs used meet norms laid down for their suitability consistently. It may also be necessary to install equipment for monitoring emissions of heavy metals.

Designing Green Cement Plants
http://dx.doi.org/10.1016/B978-0-12-803420-0.00015-9
115

(7)   Since in most cases the extent of AF used is between 20 to 30% of the main fuel, the tonnage required is small, and storage and equipment for processing would necessarily be of small capacity.

Hence capital investment would be small.

If AF is regularly available, then maintenance stocks can be reduced.

These would be large if the CC has to provide storage and processing facilities on its premises.

8.1.1   Taken as whole, capital costs for introducing an AF System in a running plant or for including it at the design stage in a new plant would be a small percentage of overall capital costs.

As mentioned above, capital costs would be even less when the AF is available in ready-to-use form. The CC would have to pay a higher price for the AF, which would be reflected in the production costs.

8.1.2 Data on capital costs

Table 4.8.1 furnishes data available for some installations in India. It is to be taken as indicative only.

8.2 Benefits of using AF

The benefits accrued from using AF are measured in three planes. On one plane, benefits are measured in terms of reduction in GHG emissions as a result of AF. Carbon-neutral AFs contribute to the reduction of GHG emissions to a much greater extent than other types of AFs. See Figure 4.8.1.

On the second plane, benefits are measured in terms of fossil fuels saved. Savings are in direct proportion to the quantum of substitution.

Thirdly, benefits are measured in fuel cost savings and are proportional to the differential costs of fossil and AF calculated on a cost per kilocalorie as fired basis.

Cement companies would more interested in derived economic benefits.

8.2.1   If AF use results in a slight increase in specific fuel consumption and in specific gas volumes, this should also be taken into consideration. In short, costs of production as a whole should be taken into account.

Table 4.8.1 Brief Data on Some of the AFR Installations in India

Sr. No.	Company	Alternate Fuels Used	% Used	Fed into	Investment (Rs. Lakhs)	Payback (Years)	Problems	Impact
1	A	A variety of industrial wastes and biomass			400			Successful no problems in emission gases
2	B	Plastic waste		Kiln				Successful
3	C	Sludge from common effluent treatment plant hazardous waste	10 t/day	Calciner	Minimal		Jamming	Successful
4	D	Locally available rice husk					Jamming	Successful
5	E	Pet coke, coffee husk, cashew, ground nutshells, etc.		Calciner			Storage and extraction; low bulk density, moisture	Savings Rs. 8 crores per year
6	F	Sludge in LSHS, fuel, hazardous waste						Rs. 2.8 lakh per year
7	G	Tire chips, husk, rubber dust, processed msw		Calciner	Rs. 200 lakhs		Jamming, surges	
8	H	Pharmaceutical, waste, paint sludge, etp sludge	5–10%		Rs. 100 lakhs			

Continued

.Table 4.8.1 Brief Data on Some of the AFR Installations in India—cont'd

Sr. No.	Company	Alternate Fuels Used	% Used	Fed into	Investment (Rs. Lakhs)	Payback (Years)	Problems	Impact
9	J	Agro and industrial wastes, tire chips, msw, paint sludge locally available			Rs. 250 lakhs		Pollution board clearance	Plastic and wood shredding machine
10	K	Waste oil from, gearboxes, hazardous wastes, categories 5.1, 5.3, 3.3	400 l per hour		Rs. 4 lakhs	18 months	Incomplete combustion of waste oil	Coal saved 2.5-3 tph
11	L	Municipal waste, agricultural waste, waste-derived fuel, tire chips	Designed for up to 15%					Walking floor storage 7, extraction system
12	M	Biomass, scrap tire, animal waste, etp, bio solid sludge, msw	Total up to 8%, tires 3%, husk 3.50%		Rs. 85 lakhs	4 years		
13	N	High calorific value, organic residue					Difficult to flow high sp. gravity	Coal saved 525 tons/year, waste disposal cost less by 50%
14	O	Textile waste, rejected shampoo		Kiln riser duct				Up to 20% fuel can be substituted by waste fuel

Source: Case study manual on use of alternate fuels and raw material by CII.

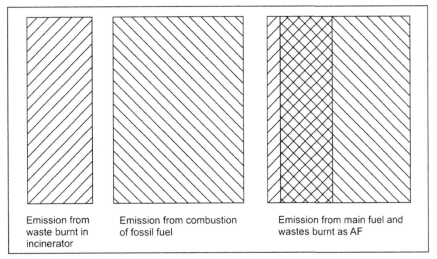

Emission from waste burnt in incinerator

Emission from combustion of fossil fuel

Emission from main fuel and wastes burnt as AF

Figure 4.8.1 GHG emissions savings with AF use. *(Source: Alternate fuels in cement manufacture. Technical and Environmental Review, a CEMBUREAU Presentation).*

8.3 Payback period

Payback period is worked out in the usual manner (investment costs/annual profit). This will vary greatly for the same AF from country to country and for different AFs, depending on the capital investment made.

Table 4.8.1 shows that payback periods range between 6 and 24 months.

CCs that have actually installed systems for AFs are in a better position to furnish more accurate information on this.

8.4 Intangible benefits

Besides the benefits in terms of reduction in GHG emissions and savings in fossil fuel costs mentioned above there are other significant though indirect benefits.

(1) In making blended cements, fly ash and slag that are wastes of their respective industries are used. Enormous problems with their disposal and dumping are solved.

(2) In the same way, use of AFs resolves to some extent problems of disposing of AFs that are wastes for their respective industries. Problems of burnable wastes are more severe than problems of inert waste.

Thus, use of AF saves on land for dumping and the nuisance of burning them and prevents pollution of rivers and streams where they may be dumped otherwise.

These are benefits for the whole community rather than for the industries producing these wastes. These are benefits to the environment and to the community as a whole.

(3) Cities are growing fast in numbers as is population across the world. This creates serious problems with waste (MSW) disposal and treatment. MSW is a significant type of AF. Hence its use on an increasing scale greatly helps solve urban problems.

8.5 When making decisions about AF, these wider aspects should also be taken into account.

Fortunately, experience shows that this is a win–win situation, as is the making of blended cements. This is evidenced by the fact that the cement industry in Europe has replaced fossil fuel by AF by as much as 40% and is marching toward "zero fuel costs".

CHAPTER 9

Alternative Raw Materials

9.1. Alternative raw materials

As for Alternate Fuel (AF), there are several possible substitutes for conventional elements of raw materials and additives that go into making cement clinker.

Among them are:
Clay minerals to substitute for/replace alumina
Lime sludges to substitute for lime
Red mud to substitute for iron

9.1.1. There are a great many possibilities and new possibilities will present themselves in the future. These should be treated as opportunities to save natural resources.

9.2. Using ARs (alternative raw materials)

ARs can be easily added in a running plant in the raw material grinding section. All it takes is the addition of an extra hopper and a metering system.

However when the AR is very different from the material it is to substitute for (for example: lime sludge is wet, and red mud is also in sludge form) pretreatment is necessary. Sludges have to be dried before adding them to the usual components in a raw mill.

9.3. Using **ARs** is different than making composite cements, in that the addition is in pre-clinkering stage.

A brief list of likely **ARs** is attached. See Table 4.9.1.

Designing Green Cement Plants
http://dx.doi.org/10.1016/B978-0-12-803420-0.00016-0

Table 4.9.1 Alternative (Substitute) Raw Materials and Additives (ARs)

Sr. No.	Main Raw Materials & Additives	Possible Substitutes, Alternative Materials	Source of Substitutes
1	Clay minerals for alumina	Coating residues	Foundries
		Aluminum recycling sludge	Aluminum industry
2	Limestone/$CaCO_3$	Industrial lime	Neutralization process
		Lime sludge	Sewage treatment Fertilizer plants
3	Silicates/SiO_2	Foundry sand	Foundries
		Contaminated soil	Soil remediation
4	Iron oxide/Fe_2O_3	Roasted pyrite	Metal surface treatment
		Red mud	Metal industry Industrial waste water treatment
5	Si-Al-Ca-Fe	FI ashes	Incinerator
		Crushed sand	Foundries
6	Sulfur	Gypsum from gas Desulfurization	Incineration Neutralization process
		Chemical gypsum	
7	Fluorine	$CaF2$ filter sludge	Aluminum industry

Source: CII: Case Study Manual on Use of AFR.

CHAPTER 10

Quality Control

10.1. Need for rigorous quality control

When firing AFs, quality control of the AF received and clinker produced assumes great importance. This is because of a great many variables due to AF coming from a great number of sources.

10.2. It is known that a number of AFs with good heat value and other desirable characteristics also contain high chlorine and dioxins and heavy metals which cause operational problems.

While a great many elements get combined with clinker in the burning process without causing any harm to its quality, excess chlorine forms compounds in riser duct and in the last two stages of preheating, which cause choking.

> **10.2.1.** Heavy metals are objectionable elements in kiln exhaust gas emissions. Therefore limited content of such elements is prescribed in the composition of exhaust gases. See Table 4.1.1 in Chapter 1.

> **10.2.2.** It is therefore necessary for a cement company to have equipment for monitoring these parameters, over and above those they monitor when producing clinker with fossil fuels.

10.3. Quality checks at source and during processing stages

Because waste fuels are literally wastes at their source, generators are unlikely to worry about the quality of the waste. Therefore, a watchful eye must be kept on waste fuels wherever and whenever they are collected. It does not matter who collects them, cement companies or ISPs.

> **10.3.1.** Various stages of inspection and checks have been shown graphically in Figs. 4.5.1–4.5.3 in Chapter 5.

>> **1.** Inspection and testing at source to determine basic criteria for suitability as AF, such as:
>> heat value
>> moisture

Designing Green Cement Plants
http://dx.doi.org/10.1016/B978-0-12-803420-0.00017-2

pH value

chlorine

2. More detailed testing of representative samples, called **fingerprinting**, is carried out in labs of the CC or the ISP to confirm suitability, such as:

physical properties

grain size analysis

moisture

odor/obnoxious gases

reactivity

pH value

ash

useful calorific value

elements like chlorine, sulfur, heavy metals

When there is more than one source of AF, fingerprinting should be done for representative samples from each source.

10.4. Fingerprinting indicates the processing required for different AFs. This has already been discussed in Chapter 4, Table 4.4.1.

The most suitable processing is then carried out by installing equipment suitable for the purpose.

10.4.1. Sometimes blending is necessary to obtain uniformity of quality. In that case, quality checks should be carried out in the same manner at different sources.

10.5. If processing is done by an ISP, quality checks must be carried out at all stages before dispatch.

If processing is done by the CC, it must devise a procedure for periodic tests right up to the point the AF is fired into the kiln and/or calciner.

10.5.1. This may sound too rigorous in the beginning, but it is necessary when beginning to use AFs. It becomes routine when good procedures are laid out at the start.

10.6. Tests on product

These are the same as those routinely done in the production of clinker.

However if certain problematic elements are known to be present in the AF being used, special checks should be regularly carried out to ensure that clinker quality is acceptable by established norms and that harmful emissions are within limits.

10.6.1. The laboratory of the CC has to be equipped with apparatus to carry out these checks.

10.7. Emissions

Continuous monitoring of exhaust gases is required at least initially, and most certainly during pilot trials necessary to obtain statutory clearances.

This part of quality control falls within the domain of the CC.

10.8. Equipment in laboratories
Laboratories of the CC have to be equipped with the following:
Brook field viscometer
Ion analyzer
Calorimeter
CHNS analyzer
Absorption spectrophotometer

10.8.1. ISPs may require vans fitted with necessary testing apparatus as they go from place to place identifying AF.

10.9. In India GEPIL is one such ISP. GEPIL is equipped with various types of the testing facilities mentioned above and is in a position to process various types of AF and make them available to cement and other industries in ready-to-use form. For AF use to increase there is an urgent need for many such companies.

CHAPTER 11

Procedure for Beginning to Fire AF on a Continuous Basis

11.1 While alternate fuels (AF) of different types has been used and hence its viability has been established convincingly, any cement company (CC) that proposes to use AF from the design stage in a new plant or in an existing plant must follow certain procedures to satisfy itself and the statutory authorities that the AF being used has no adverse effect on either the quality of clinker produced or on the emissions.

11.2 Pilot plant trials

For this reason it is best to carry out pilot plant trials whenever possible. Such trials can be conducted on the premises of research laboratories or institutions that have expertise in the field.

11.2.1 Pilot plant trials also reveal processing required to impact the quality of clinker and the exhaust gases.

They also indicate further steps required, such as:

1. design modification of the calciner to control NO_x and sulfur
2. whether bypass is required to keep emissions of chlorine and heavy metals down
3. the extent to which AF can replace the main fuel
4. operational impact, such as the increase in sp. fuel consumption and specific gas volumes. The system for installing the AF firing can be designed in accordance with the findings of the pilot plant trials.

11.3 The system to be installed within the plant will include arrangements for receiving and storing AF, its extraction from storage, and conveying it to the feeding system.

11.3.1 If processing is to be done within the plant then equipment is required for processing and for feeding the processed AF to

Designing Green Cement Plants
http://dx.doi.org/10.1016/B978-0-12-803420-0.00018-4

the feeding system. This will in all probability be installed outside the plant.

11.4 A full-fledged trial of firing AF will be performed, beginning with a small percentage and gradually increasing the percentage in steps, monitoring all the while operating and quality parameters.

11.5 Sanction from authorizing bodies

Pollution control boards (PCBs) in respective countries prescribe procedures for conducting the trials and operational parameters to be observed under their supervision. CCs have to be equipped for conducting the trials and for recording all observations to satisfy PCBs and concerned standards institutions (SIs).

PCBs and/or SIs will then give necessary clearances to continue to fire AF.

11.5.1 It is desirable that representatives of the PCB and/or SI witness either the pilot plant or plant trials.

11.6 Viability and other aspects must also be checked by the CC. The CC also works out the impact on emissions (the reduction in CO_2 per ton of clinker produced as a result of firing AF).

11.7 Thus the actual use of AF at an optimum level on a continuous basis may be a long, drawn-out process, as outside agencies are involved.

11.8 If the CC proposes to use an intermediate service provider (ISP) to procure preprocessed ready-to-use AF, it must negotiate with the ISP terms such as tolerances in quality, continuity of supply and stocks to be maintained in case of emergencies.

11.8.1 The procedure for testing the AF before it is delivered to the plant should also be established, so that there is no rejection.

CHAPTER 12

Problems with Using AFs

12.1 Cement companies proposing to introduce use of AF should acquaint themselves with problems that they may have to face in their use. Thereby they can equip themselves with means to overcome them.

12.1.1 Likely problems
1. availability on a continuous basis in required quantities
2. consistency in properties like calorific value, moisture, density, etc.
3. preprocessing required on many kinds of AF to make them ready to use (this differs from waste to waste)
4. many wastes with useful heat value are hazardous to store and handle and need special care.

12.2 Problems of logistics

Cement plants are primarily located near sources of limestone deposits.

Fuels presently in use, that is coal, oil, etc., are brought to the cement plants.

Compared to limestone deposits, locations of coal and oil sources are few. Hence even in the case of fossil fuels the logistics of bringing fuels to cement plants poses problems.

12.2.1 Sources of AF are too many
1. Biomass fuels (agricultural wastes) are available over large areas of a country but availability in any one place is not sufficient to meet the needs of the present–day-size plants. Further, some of them are seasonal.
2. Industrial wastes are available in industrialized parts of the country. Different types are available in different parts of a country. Again availability in desired quantity of any one type of AF is uncertain.
3. MSW is available from large cities, sometimes far removed from cement plants.

Designing Green Cement Plants
http://dx.doi.org/10.1016/B978-0-12-803420-0.00019-6

Unless infrastructure is developed simultaneously to collect wastes from their sources of generation and bring it to the doorsteps of cement plants, it will time for the use of AF to gather momentum.

12.3 Checks on quality—acceptance and rejection

In the case of AFs, close watch must be kept on pertinent properties like calorific value, moisture, and impurities, so that fuel once collected does not need to be rejected. This is particularly true of waste-derived fuels. This limitation naturally restricts their availability.

12.4 Processing required for AFs

Most biofuels and waste oils and sludges have to undergo a great many operations, like cleaning, drying, and size reduction, before they can be used as fuels.

Some are made into briquettes and some are pulverized. A good many types are best turned into gas for use.

This processing requires considerable capital investment, and the free-of-cost fuels (because they are wastes) are no longer free.

The higher the differential between the cost of coal and that of AF, expressed in rs per kcal, the greater the incentive to use AF.

However it should not be forgotten that even coal has to undergo processing like blending, crushing, drying, and grinding to make it ready for use.

12.5 Operational difficulties

On the operational side there may be difficulties feeding fuel in predetermined proportions, because when compared to coal AF has more fluctuation of properties like density, calorific value, etc.

A closer watch needs to be kept on undesirable emissions in exhaust gases.

12.5.1 Biomass

Biomass fuels tend to decompose if stored for long when wet. They can also emit methane gas, which is combustible; hence, there may be fire hazards as well. Therefore biomass fuels need to be dried on receipt if wet. In storage they also lose calorific value due to loss of volatiles.

The sheer volume of biomass due to low bulk density makes it unmanageable.

Efficiency of energy conversion is low for biomass fuels as compared to fossil fuels.

12.6 The problems mentioned above have already been overcome in European countries, which use AF to the extent of 40%. Hence these problems should not deter CCs from beginning to use AF.

12.7 Research

There is urgent need for concentrated research on waste fuels available and how they can be used.

12.8 Infrastructure

There is urgent need to develop infrastructure to make ready-to-use AF available to cement and other industries. There is great scope for this in large countries like India which have made a start in using AF.

12.9 AF may be available free or at negligible cost initially, but as more and more industries begin to use them availability will decrease and they may come at a price. Over time they may have to be brought over longer and longer distances.

CHAPTER 13

Recommendations and Conclusions

13.1. Feasibility

The feasibility of using diverse types of AF, including biomass, RDF, industrial wastes, and MSW, has been proven in many countries all over the world.

Countries using AF include the advanced and developed countries of Europe and also less developed countries in Africa and Asia.

13.1.1. The extent of substitution of fossil fuels by AF varies from country to country. In European countries substitution to the extent of 40% has been achieved. In countries like India, where firing AF is in early stages, substitution varies between 5% and 10%.

13.2. Technology

Technology for firing different types of AF has also been developed.

Various steps required to prepare various types of AF from their "as is where is" form to ready-to-use form are also known. Machinery required for this purpose, ranging from crushers and shredders to centrifuges, dryers and even gasifiers is available or can be designed to meet specific requirements. Conveying equipment and feeders of various types designed to suit different AFs are also available.

Special machinery like Hot Disc and Gunnax for firing whole tires in kilns are also now available.

13.3. Quality checks

Greater control must be exercised over sampling and testing AF at the collection stage because a great number and types of AF are, after all, wastes. In the case of using wastes like plastics there is more chlorine, and there is the possibility of finding dioxins and heavy metals in exhaust gases. This needs to be monitored closely.

Designing Green Cement Plants
http://dx.doi.org/10.1016/B978-0-12-803420-0.00020-2

13.3.1. Hazardous wastes and biomass

Some wastes are hazardous to store, handle, and transport. Particular care is required when using them as AF. Biomass decomposes if stored for long. It also loses its heat value. There is also a possibility of generation of methane.

All of the above may seem forbidding and may deter cement manufacturers from beginning to use AFs. But, as has been mentioned, solutions have evolved to address most of them.

13.4. Benefits

These have been discussed in sufficient detail. Besides direct benefits in terms of energy costs, there are other benefits to the community, to the country, and to the world as a whole.

Use of AF:

1. saves nonrenewable fossil fuels
2. reduces GHG emissions and contributes to slowing of global warming
3. reduces waste disposal problems of communities, cities, and industries

13.5. Principal problems

Availability of AF in ready-to-use form on continuous basis and at the scale required is likely to be the principal problem for the cement and other industries desirous of using AF because of the lack of infrastructure in a great many countries.

It is desirable that such infrastructure is developed simultaneously. This will speed up the use of AF.

13.6. Technical assistance

New entrants to the fraternity of AF users need guidance at various stages in order to design the AF system best suited to their plants. Existing consultancy companies need to acquire know-how in this area.

13.7. Outsourcing

Cement companies find it difficult to take on the additional responsibility of collecting AFs on an "as is where is" basis.

Companies specialized in handling and processing different types of AF should become available in different parts of the country to serve users in those regions. These are **intermediate service providers (ISPs).**

Without the simultaneous growth of this infrastructure, increased use of AFs will be slow.

13.8. Research

Considerable research is still needed regarding selection of AF, determining the proportions to be used with the main fuel, and the impact on quality of clinker, emissions, gas volumes, and specific fuel consumption.

Rather than individual plants, this is best undertaken by national institutions (like the Central Fuel Research Institute and NCCBM in India, for example), setting up new sections to carry out research. Research results should be available to the whole country.

13.9. Incentives

Maybe it would initially be necessary to offer financial incentives to companies to encourage them begin to use AF.

Carbon credits are available to companies who reduce GHG emissions.

13.10. Conclusions

There is undoubtedly a great future for the use of AF. Present difficulties are the teething troubles that all new ventures have to overcome. There is reluctance to use AF because of teething troubles. But teething problems can be overcome only by greater use of AF. In a way this is kind of a "chicken or the egg" situation.

Hopefully the use of AFs and the infrastructure needed will grow together, complementing one another and benefiting the country and the community as a whole.

RECOMMENDED READING

1 Confederation of Indian Industries (CII), Sorabjee Green Business Centre - Case Study Manual on Alternate Fuels & Raw Materials (AF&R) Utilization in Indian Cement Industry 2011

2 Science Direct Perspectives and Limits for Cement Kilns as Destination for RDF by G Genon & E Brizio 2007

3 Centre for Sustainable Development India Co-incineration of MSW in Cement Industry by : Axel Seeman 2007

4 Use of AF in Indian Cement Industry by : S R Asthana & R K Patil CEPD Division, National Chemical Laboratory, Pune, 2006

5 Cembureau Presentation 1997 Alternative Fuels in Cement Manufacture - Technical & Environmental Review

6 Melbourne Water - ANZBP Hobart Roadshow 2011 Beneficial Uses of Bio solids in Cement Production

7 United Nations Development Programme (UNDP) 2009 Biomass Energy for Cement Production - Opportunities in Ethiopia

8 Cembureau Environmental Benefits of Using AF in Cement Production- A Life Cycle Approach

9 Use of AF in Cement Manufacture - Analysis of Fuel Characteristics & Feasibility for Use in Chinese Cement Sector by : Ernest Orlando Lawrence 2008 - Berkeley National Laboratories

10 PREGA Phillipines Utilsation of AF in Cement Manufacture - A feasibility Study Report 2005

11 Waste Preparation Facility Presentation by Priyesh Bhatti of GEPIL (Gujerat Enviro Protection & Infrastructure Ltd)

12 Efforts for Earth - Brochure of GEPIL.

SECTION 5

Waste Heat Recovery

Contents

List of Flow Charts and Layouts

List of Tables & Annexures

List of Plates

CHAPTER 1

Introduction

1.1 Green cement and WHR

WHR is an important aspect of green cement. WHR directly reduces operational costs by converting waste heat into electricity, reduces consumption of nonrenewable fossil fuels, and reduces emission of greenhouse gases (GHGs).

    1.1.1 Power generated by using waste heat from kiln and clinker cooler exhaust gases can be used within the plant or fed to the grid. Thus there is a net savings in power taken from the grid on a regular basis.

1.2 The savings in fossil fuel as a result of less power generation at the grid results in reduced GHG emissions per ton of cement produced.

For these reasons WHR is almost always considered an integral part of a new cement plant. It can also be considered for existing cement plants.

1.3 There are many dimensions to the proposition of WHR. They need to be considered while designing and engineering WHR systems.

1.4 Feasibility and viability of cogeneration or WHR

First, it is necessary to establish the feasibility and viability of installing a WHRS for each and every situation as there are many variables.

Variables that affect the viability are:

1. size of the plant
2. net heat available from kiln and cooler waste gases, which in turn dictates the system of recovery from among the three well-known systems:
   standard steam Rankine cycle
   organic Rankine cycle
   Kalina cycle

Designing Green Cement Plants
http://dx.doi.org/10.1016/B978-0-12-803420-0.00005-6

3. overall power situation

This includes three entities:

1. grid power as main source of power
2. captive power as standby power in times of power scarcity
3. power that can be generated from WHR

An optimum solution involving minimum capital investment needs to be determined in each case.

1.5 Various aspects mentioned above, beginning with the potential for generating power from WHR, are dealt with in the following chapters in sequence.

Different systems are explained with the help of system flow charts and discussion of salient features.

The same system can be installed in different ways:

1. separate systems for kiln and cooler gases
2. common system for the two.

1.6 If a captive power plant is also to be included in the project, WHR would have to be integrated between it and the cement plant.

1.7 Naturally capital costs depend on final choices regarding the system, capacity and auxiliaries that have to be added to support WHR.

1.8 Payback period needs to be worked out as that will be the ultimate yardstick for the viability of the project.

1.9 WHR systems have been installed the world over and their contribution to fossil fuel savings and reduction in GHG emissions is beyond any doubt. It is the investment costs expressed in terms of Rs/MW compared to similar costs for a new thermal power station (TPS) that are the deciding factor in most cases.

1.10 Potential for WHR

It is therefore necessary to establish the potential for installing WHR in a new plant or in an existing plant as a first step.

This is easily done by ascertaining

1. the heat content in waste gases of the kiln-preheater system and the cooler vent.

2. requirements of drying (a) raw materials, (b) coal, (c) slag, as the first preference would naturally be given to this need before using waste heat for generating power.

The net waste heat available for WHR is the difference between the two above.

1.11 Special conditions prevailing in cement plants vis-a-vis WHR

1. waste gas volumes and temperatures vary, albeit in a small range, even in normal operation.

Cooler vent volumes vary by as much as 30% during kiln upset conditions, when temperatures are almost doubled.

2. requirements of drying raw materials and coal vary from season to season.

First priority would always be given to drying them.

3. dust loads are high, at 70-100 g/nm^3 and 10-15 g/nm^3 respectively for preheater exhaust and cooler vent.

Dust in preheater exhaust is sticky. Dust in a cooler vent is abrasive.

4. high-efficiency cement plants need to be operated at close to full load conditions.

1.11.1 Temperatures of preheater gases cannot be reduced to a level below those required for drying raw materials and coal.

1.11.2 A WHR system should be so installed that it do not interfere with the normal operation of the kiln. It should therefore be installed in "bypass" mode.

For satisfactory operation of WHR, a plant also must operate between 75% and 80% of its full capacity.

1.12 Choice of systems for cogeneration

The temperature difference and heat content decide which of the three systems mentioned in paragraph 1.4 is suitable in any given case.

There are presently three established systems

1. Rankine cycle using water as the medium for generating steam.

2. Kalina cycle. which uses a water and ammonia mixture as the medium.

3. organic Rankine cycle, which uses thermal fluids as the medium.

Details of these systems with flow charts and salient features are furnished in Chapter 3.

1.12.1 Capital costs and operational efficiencies of the three systems vary and hence viabilities and payback periods can also vary. The selection of the system should be done carefully, with the help of consultants and equipment suppliers. There can be many alternatives and these should be considered.

1.13 Options

There are several ways in which a WHRS can be installed in cement plants in addition to the types of WHRS mentioned above.

Case 1   Four- to five-stage preheater and grate cooler, with preheater gases used for drying raw materials and coal.

The most suitable system would be the common Rankine cycle, depending on the size of the plant.

Further options would be:

1. separate systems for preheater gases and cooler gases
2. common system using preheater and cooler gases.

One may also consider whether WHRB is installed before or after the dust collectors.

System design is also influenced when alternative fuels are used as gas volumes may be different.

Case 2   Six-stage preheater and latest high-efficiency grate cooler, with preheater gases to be used for drying raw materials and coal

1. Common or separate systems. In all probability a common system will select itself.
2. Since temperatures of waste gases are low, in all probability either a Kalina or an Ormat system would be used.
3. If it is proposed that the plant has a captive TPS, one may also consider whether the waste gases could be used in the

proposed TPS, and a TPS with correspondingly higher capacity could be installed.

For example, let the requirement of the plant be 50 MW, with a proposed captive TPS corresponding of 30%.

Then the TPS would have a rating of 15 MW.

A WHRS for a 10,000-tpd kiln can have a capacity of ~12.5 MW.

at 30 kwh/t.

1.13.1 The company can therefore decide, in addition to 50 MW of grid power,

1. to install a WHRB of 12.5 MW and a TPS of 15 MW, with grid power for full capacity 50 MW
2. to install a 30-MW TPS in stages, but using waste heat from cement plant, thus requiring less coal to be fired, avoiding a separate generating station

1.13.2 To make a final decision, it is necessary to obtain the capital costs and operating costs in each case.

1.14 Steps to be taken in installing a WHRS

Once the decision is made, the next step is detailed engineering.

This would involve firming up specifications of the TPS and WHRB and creating system flow charts to identify all the items of main machinery and auxiliaries to be included, identifying requirements of the plant as a whole (including the cement plant, WHRB and TPS) for power, water, and fuel,—and in particular special equipment specific to power generation like water circulation equipment, condensers, cooling towers, and the like.

Consultants should work out cost estimates and also prepare preliminary layouts of the TPS and WHRB to be installed in the plant. It is then possible to know the total area of land required.

1.14.1 If the WHRB is to be installed in a running plant, then it has to be fitted in the existing plant layout.

Additions would be:

- Water for WHR systems and equipment for its handling, treatment, circulation and cooling
- Cooling towers
- Condenses for steam
- Turbines and generators

In effect, additions would be similar to the corresponding equipment in the main TPS (when there is one) and electrical substation.

1.15 These various aspects will be dealt with in the following chapters.

CHAPTER 2

Waste Heat Available for Generating Power

2.1 The first step in planning to install a waste heat recovery system (WHRS) is to work out the net available waste heat that can be used to generate power.

2.1.1 In a cement plant, waste heat is available mainly from kiln exhaust gases and vent air from the clinker cooler.

Waste heat is also available from the exhaust of diesel engines of DG Set. However, that would be a small proportion of the waste heat available from the kiln and cooler system and hence is ignored here.

2.2 Waste heat available from kiln exhaust gases

The heat content of kiln-preheater exhaust gases depends on their volume and temperature.

The lower the volume and temperature, the smaller the heat content.

In dry process kilns with preheaters, the number of preheater stages is a pointer to specific fuel consumption and hence specific gas volume and the temperature of gas at the preheater exit.

For example (for illustrative purpose only):

Preheater stages		4	5	6
Fuel efficiency	kcal/kg	800	750	700
Exhaust gas Clinker	nm^3/kg	1.5	1.4	1.3
Temperature of gas (°C)		350	300	270

Assuming a specific heat of exhaust gas of 0.3 kcal/nm^3, the heat content of exhaust gases in kcal/kg of clinker (ambient at 30 °C) would be

Stages of preheater	4	5	6
kcal/kg clinker	**144**	**113**	**94**

Table 5.2.1 shows specific gas volumes for different specific fuel consumptions and calorific values of coal.

Designing Green Cement Plants
http://dx.doi.org/10.1016/B978-0-12-803420-0.00022-6

Table 5.2.1 Calculation of Specific Gas Volumes from Preheater

Sr. No.	Item	Unit			
1	Base assumptions				
	Preheater stages	No	4, 5, 6		
	Fuel		Coal		
	Calorific value	kcal/kg	4000–6000		
	Altitude		Sea level		
	Ambient temp.	°C	30		
	Total excess air	%	25		
	Excess air for combustion	%	10		
	Sp. fuel consumption	kcal/kg clinker	700–800		
	Sp. heat of air/gas	kcal/nm^3	0.3		
2	Air for combustion	nm^3/kg coal	**Calorific value of coal kcal/kg**		
			4000	5000	6000
3	Air for combustion	nm^3/kg coal	4.54	5.55	6.56
	Products of combustion	nm^3/kg coal	5.21	6.1	7
	Products of combustion with 25% excess air	nm^3/kg coal	6.35	7.49	8.64
4	Coal per kg clinker		**Sp. fuel consumption kcal/kg clinker**		
			700	750	800
		Calorific value coal kcal/kg	**Coal in kg per kg clinker**		
		4000	0.18	0.19	0.2
		5000	0.14	0.15	0.16
		6000	0.12	0.13	0.13
5	**Case 1**		**Four-stage preheater, 800 kcal/kg, 350 °C Calorific value coal kcal/kg**		
			4000	5000	6000
	Coal	kg/kg clinker	0.2	0.16	0.13
	Products of combustion	nm^3/kg coal	6.35	7.49	8.63
	Products of combustion	nm^3/kg clinker	1.27	1.20	1.12
	CO_2/kg clinker	nm^3	0.3	0.3	0.3
	Sp. gas volume	nm^3/kg clinker	**1.57**	**1.50**	**1.42**

Continued

Table 5.2.1 Calculation of Specific Gas Volumes from Preheater—cont'd

Sr. No.	Item	Unit			
6	**Case 2**		**Five-stage preheater, 750 kcal/kg, 300 °C Calorific value coal kcal/kg**		
			4000	5000	6000
	Coal	kg/kg clinker	0.19	0.15	0.13
	Products of combustion	nm^3/kg coal	6.35	7.49	8.64
	Products of combustion	nm^3/kg clinker	1.21	1.12	1.12
	CO_2 per kg clinker	nm^3	0.3	0.3	0.3
	sp. gas volume	nm^3/kg clinker	**1.51**	**1.42**	**1.42**
7	**Case 3**		**Six-stage preheater, 700 kcal/kg, 270 °C Calorific value coal kcal/kg**		
			4000	5000	6000
	Coal	kg/kg clinker	0.18	0.14	0.12
	Products of combustion	nm^3/kg coal	6.35	7.49	8.64
	Products of combustion	nm^3/kg clinker	1.14	1.05	1.04
	CO_2 per kg clinker	nm^3	0.3	0.3	0.30
	Sp. gas volume	nm^3/kg	**1.44**	**1.35**	**1.34**

2.3 Heat used in drying raw materials and coal

Coal and raw materials use dry grinding systems. Ground raw meal should not have more than 1% moisture and coal not more than 2%.

These days, generally, vertical roller mills are used to grind both raw materials and coal. Therefore minimum moisture in feed is taken as 5%, although in countries like India moisture in raw materials may be as low as 2% in dry seasons.

Average moisture in coal is taken as 10%, although it may be higher in some months. It also depends on coal sources.

2.3.1 Requirements of drying can be worked out as follows.
   Raw materials:
   - Ratio of raw meal to clinker is ~1.5. Since raw mills are designed to work for 20 h compared to 24 h for a kiln

they have a correspondingly higher capacity to consider. Therefore raw material to be dried is 1.8 kg/kg clinker.

- Water to be evaporated from 5% to 1% is equal to 0.08 kg/kg clinker.

- Heat to be supplied to evaporate 1 kg water (feed moisture 5%) is ~1500 kcal.

- Therefore heat required to dry raw materials is **120 kcal/kg clinker**

Coal:

Coal to be used depends on fuel efficiency. Let calorific value of coal be 4000 kcal/kg. Therefore coal to be used is

fuel efficiency	kcal/kg	800	750	700
coal clinker	kg/kg	0.2	0.19	0.175

A running hour factor of 1.2 is applicable in the case of coal mills also.

Therefore coal to be dried per kg clinker   0.24   0.23   0.21
heat to evaporate for 1 kg water = 1200 kcal/kg water
water to be evaporated from kg coal = 0.09 kg/kg coal

Therefore heat to be supplied for different efficiencies is as above.

Coal	kg	0.24	0.23	0.21
Heat to be supplied	kcal/kg	26	25	23
Clinker				

Therefore total heat to be supplied for drying for different

Fuel efficiencies		800	750	700
Drying raw materials	kcal/kg	120	120	120
Drying coal	Clinker	26	25	23
Total		**146**	**145**	**143**

Table 5.2.2 shows the computation of heat required for drying under different conditions.

From Tables 5.2.1 and 5.2.2 actual values can be worked out for actual operating conditions and net available waste gas heat for power generation can be arrived at.

Table 5.2.2 Calculation of Heat Required for Drying Raw Materials and Coal

Item	Unit				
Base assumptions					
Sp. heat raw materials	kcal/kg	0.24			
Coal	kcal/kg	0.24			
1 Raw materials					
Moisture in feed	%				
Raw materials		5	6		
To be dried to		1	1		
Water to be evaporated	kg/kg	0.042	0.053		
Product	Raw meal				
Heat required to evaporate water	kcal/kg	1500	1300		
Heat required/kg		**Feed moisture %**			
Raw meal		5	6		
	kcal/kg	**63**	**69**		
2 Coal					
Moisture in coal	%	10	12		
To be dried to		2	2		
Water to be evaporated	kg/kg coal	0.09	0.11		
Heat required to		10	12		
Evaporate water	kcal/kg	1300	1200		
		Feed moisture %			
		10	12		
Heat required/kg coal		**116**	**136**		
3 To convert heat required in kcal/kg clinker					
Raw materials					
Ratio raw material/clinker		1.5			
Coal		Calorific value	**Ratio coal/clinker = a**		
Ratio coal/clinker					
Sp. fuel consumption	kcal/kg	Coal kcal/kg	700	750	800
		4000	0.18	0.19	0.2
From Table 5.2.1		5000	0.14	0.15	0.16
		6000	0.12	0.13	0.13
Factor for running hours					
Kiln hours		24			
Mill hours		20			
Factor for running		1.2			
Total factor for		1.5			

Continued

Table 5.2.2 Calculation of Heat Required for Drying Raw Materials and Coal—cont'd

Item	Unit		
Raw materials			
kg raw meal/kg clinker		**1.8**	
Raw materials		**Feed moisture %**	
		5	6
Heat for raw meal	kcal/kg	63	69
Heat required for		**kcal/kg clinker**	
Raw materials	1.8	**113**	**124**
Coal		**Ratios of coal to clinker**	
		a	b
$a=$ ratio as above		**0.12**	**0.14**
$b = a \times \frac{24}{20}$		**0.13**	**0.16**
For running hours		**0.14**	**0.17**
		0.15	**0.18**
		0.16	**0.19**
		0.17	**0.20**
		0.18	**0.22**
		0.19	**0.23**
		0.2	**0.24**
Coal		Heat kcal/kg water	
		1300	1250
	Moisture %	10	12
Heat for drying	kcal/kg coal	**116**	**136**
Heat for drying	b	Heat kcal/kg clinker	
Coal	0.14	**16**	**19**
	0.15	**17**	**20**
	0.16	**19**	**22**
	0.17	**20**	**23**
	0.18	**21**	**24**
	0.19	**22**	**26**
	0.2	**23**	**27**
	0.21	**24**	**29**
	0.22	**26**	**30**
	0.23	**27**	**31**
	0.24	**28**	**33**

However, when mills are not running kiln, gases are not taken for drying in those sections, and for that period more heat is available for generating power.

The WHRB would have to be sized on the basis of gross heat available.

It can be seen that as the operational efficiency of a cement plant increases, less and less waste heat is available for generation of power.

2.4 Properties of gases available from kiln

Exhaust gases from a kiln contain mildly abrasive dust, to the extent of $\sim$70-80 g per nm^3. The higher the efficiency of the top cyclone, the smaller the dust content.

The higher the specific gas volume (meaning lower operational efficiency), the lower the dust burden in the exhaust gas.

Further, the dust can often be sticky, depending on the content of clay-like materials and their properties.

Waste heat boilers to be used for generating power should be designed to take these factors into account.

The location of the WHB with regard to the system dust collector assumes importance.

2.5 Waste heat available from vent gases from clinker cooler

Air is used to cool clinker coming out of the kiln at around 1400 °C in all types of coolers. In recent years there have been momentous developments in the design of clinker coolers. Old reciprocating grate coolers have been replaced by new concepts like cross bar coolers and walking floor coolers.

They have been developed to handle outputs as high as 10,000-15,000 tpd capacity kilns. Their efficiencies have consistently improved, which in turn has helped in achieve overall operating efficiencies of 700/650 kcal/kg clinker for the kiln system taken as a whole.

The quantum of cooling air used to achieve the same temperature of clinker at the cooler exit has been reduced progressively. It is between 2 and 2.2 nm^3/kg clinker. The specific load of the cooler expressed as tons per day per sq. meter of cooler area is presently 45-50.

2.5.1 Vent air from the cooler is the difference between cooling air and air used for combustion in the kiln and calciner. The quantity of combustion air is dependent on specific fuel consumption and the calorific value of the coal.

For example, for coal of calorific value 4000 kcal/kg and for specific fuel consumption of 700 kcal/kg, air through the kiln would be 0.9 kcal/kg clinker.

See Table 5.2.3 which shows calculation of vent air from the cooler.

Temperatures of vent air would vary depending on its quantity and the temperature to which clinker is to be cooled. Presently vent air temperatures are 200-250 °C.

2.5.2 Cooler vent contains clinker dust of the order of 10-15 g/nm^3. It is coarse and abrasive. WHB to be designed to generate power from cooler vent has to take this factor into account.

2.6 Net waste heat available for generation of power.

Total waste heat available is the sum total of heat in waste gases from preheater and in cooler vent.

Net heat available for cogeneration is the gross heat less heat required to dry raw materials and coal which of course depends on moisture in feed in raw materials and coal as explained in paragraph 2.3.1

2.6.1 Annexure 1 shows the calculation for net waste heat available for cogeneration.

2.7 To illustrate the point with an example, let us assume:

Calorific value of coal	4000 kcal/kg
Specific fuel consumption	700 kcal/kg
Cooling air	2.2 nm^3/kg clinker
Temperature of vent air	250 °C

Table 5.2.3 Calculation of Vent Air from Cooler in nm^3/kg Clinker

Item	Unit			
		1 Calculate air for combustion through kiln and calciner		
Base assumptions				
Altitude	m	Sea level		
Ambient temp.	°C	30		
		Calorific value coal, kcal/kg		
		4000	5000	6000
		Air for combustion in nm^3/kg coal		
		4.54	5.55	6.56
		With 10% excess air		
		5.0	6.1	7.22
		Leakage through hood @ 5% nm^3/kg coal		
		0.25	0.31	0.36
		Air through kiln and calciner nm^3/kg coal		
		4.74	**5.80**	**6.86**
		2 Calculate air for combustion through kiln and calciner in nm^3/kg clinker		
		Calorific value coal kcal/kg	**Sp. fuel consumption, kcal/kg clinker**	
	kg/kg		700 | 750 | 800	
			kg coal per kg clinker	
		4000	0.18 | 0.19 | 0.20	
		5000	0.14 | 0.15 | 0.16	
		6000	0.12 | 0.13 | 0.13	
Air for kiln and calciner	Sp. fuel con.	Ratio coal/ clinker	**Air through kiln and calciner, nm^3/kg coal**	
			4.74 | 5.8 | 6.86	
			Air through kiln and calciner, nm^3/kg clinker	
	700	0.18	0.9	
		0.14	0.7	
		0.12	0.6	

Continued

Table 5.2.3 Calculation of Vent Air from Cooler in nm^3/kg Clinker—cont'd

Item	Unit				
	750	0.19		1.1	
		0.15		0.9	
		0.13		0.8	
	800	0.2			1.4
		0.16			1.1
		0.13			0.9

3 Vent air from cooler, nm^3/kg clinker
vent air = cooling air—air through kiln & calciner
Cooling air nm^3/kg clinker

Air through kiln calciner, nm^3/kg clinker	2	2.2	2.4
	vent air, nm^3/kg clinker		
0.6	1.4	1.6	1.8
0.7	1.3	1.5	1.7
0.8	1.2	1.4	1.6
0.9	1.1	1.3	1.5
1	1	1.2	1.4
1.1	0.9	1.1	1.3
1.2	0.8	1	1.2
1.3	0.7	0.9	1.1
1.4	0.6	0.8	1
1.5	0.5	0.7	0.9

Using Tables 5.2.1–5.2.4,

1 Heat in waste gases from preheater   94 kcal/kg clinker
2 Heat in cooler vent   53 kcal/kg clinker
   Total   **147 kcal/kg clinker**
   Less heat is required for drying assuming
   moisture of raw materials to be 2% and
   that of coal 10%. Heat required for
   drying would be:
   for drying raw materials   ~42 kcal/kg clinker
   for drying coal   23 kcal/kg clinker
   Total   **65 kcal/kg clinker**
   Net heat available for cogeneration   **82 kcal/kg clinker.**

Table 5.2.4 Heat content of waste gases for different inlet and outlet temperatures

Item	Unit	Quantity	1 Kiln waste gases		
Sp. heat of gas/air	kcal/nm³	0.3			
Temperature of waste gas **Temp. at inlet °C**			270	300	350
Temp. at outlet °C			30	30	30
Useful heat content kcal/nm³		a	72	81	96
Temp. at outlet °C			90	90	90
Useful heat content kcal/nm³		b	54	63	78
Temp. at outlet °C			200	200	200
Useful heat content kcal/nm³		c	21	30	45
2 Cooler vent **Temp. at inlet °C**			200	250	300
Temp. at outlet °C			90	90	90

Continued

Table 5.2.4 Heat content of waste gases for different inlet and outlet temperatures—cont'd

Item	Unit	Quantity			
		d	33	48	63
Useful heat content	kcal/nm³				
3 Heat content for different volumes and temperatures of kiln preheater gases					
			Heat content kcal per nm³ vide c above		
Gas volume nm³/kg clinker			21	30	45
			Heat content kcal per kg clinker		
		1	21	30	45
		1.1	23.1	33	49.5
		1.2	25.2	36	54
		1.3	27.3	39	58.5
		1.4	29.4	42	63
		1.5	31.5	45	67.5
		1.6	33.6	48	72
Gas volume nm³/kg clinker			**Heat content kcal per nm³ vide b above**		
			54	63	78
			Heat content kcal per kg clinker		
		1	54	63	78
		1.1	59.4	69.3	85.8
		1.2	64.8	75.6	93.6
		1.3	70.2	81.9	101.4
		1.4	75.6	88.2	109.2
		1.5	81	94.5	117
		1.6	86.4	100.8	124.8
Gas volume nm³/kg clinker			**Heat content kcal per nm³ vide a above**		
			72	81	96

Heat content kcal per kg clinker

	72	81	96
1	72	81	96
1.1	79.2	89.1	105.6
1.2	86.4	97.2	115.2
1.3	93.6	105.3	124.8
1.4	100.8	113.4	134.4
1.5	108	121.5	144
1.6	115.2	129.6	153.6

4 Heat content in cooler vent air

Heat content in kcal per nm³ vide d above

Vent air vol nm³/kg clinker	33	48	63
0.5	33	48	63
0.6	16.5	24	31.5
0.7	19.8	28.8	37.8
0.8	23.1	33.6	44.1
0.9	26.4	38.4	50.4
1	29.7	43.2	56.7
1.1	33	48	63
1.2	36.3	52.8	69.3
1.3	39.6	57.6	75.6
1.4	42.9	62.4	81.9
1.5	46.2	67.2	88.2
1.6	49.5	72	94.5
1.7	52.8	76.8	100.8
1.8	56.1	81.6	107.1
	59.4	86.4	113.4

This quantum would decrease considerably if the moisture to be dried in raw materials is higher.

2.8 In some specific cases it may be necessary to dry slag brought in to make slag cement. In such cases cooler vent air may have to be used.

2.9 One particular feature of kiln operation that considerably influences the temperature of vent air and its quantum expressed in nm^3/kg clinker is the operation of kiln under upset conditions.

The WHB for using cooler vent air has to take into account this condition as it may last for 3-4 h at a time.

2.10 Selection of waste heat boiler

The quantum of waste heat available and the temperature at which it is available more or less decide the systems to be used for cogeneration.

Apart from the system choice, there are also several choices regarding how to fit the system.

These aspects are dealt with in the following chapters.

ANNEXURE 1

Example of Calculating Net Heat Available from Waste Gases for Cogeneration of Power

Cement Plant Data		Common WHR for Kiln and Cooler	
		Operating Conditions	
Item	Unit	1	2
Altitude	m	Sea level	Sea level
Ambient temp.	°C	30	30
Fuel		Coal	Coal
Useful calorific value	kcal/kg	5000	5000
Moisture in wet coal	%	10	10
To be dried to	%	2	2
Mill running hours	no	20	20
Raw materials to			
Clinker	Ratio	1.5	1.5
Moisture in wet			
Raw materials	%	5,3	5,3
To be dried to	%	1	1
Mill running hours	no	20	20
Preheater stages	no	**6**	**4**
Sp. fuel consumption	kcal/kg	**700**	**800**
Temp. of gases leaving preheater	°C	270	350
Sp. heat of gases and air	kcal/nm^3	0.3	0.3
Heat required to evaporate water in raw materials	kcal/kg	1500/1800	1500/1800
Heat required to evaporate water in coal	kcal/kg	1300	1300
Cooling air	nm^3/kg clinker	2.2	2.4
Temp. of cooler vent air	°C	250	250
Cooler vent cooled to	°C	90	90
Preheater gases in raw mill and coal mill cooled to	°C	90	90
CO$_2$ released per kg clinker	nm^3/kg	0.3	0.3
Sp. gas volume from preheater	nm^3/kg clinker	**1.35**	**1.5**
Vent air from cooler	nm^3/kg clinker	**1.4**	**1.5**
Heat content of preheater gases	kcal/kg clinker	**72.8**	**117**

Continued

Example of Calculating Net Heat Available from Waste Gases for Cogeneration of Power—cont'd

Cement Plant Data		Common WHR for Kiln and Cooler	
		Operating Conditions	
Item	Unit	1	2
heat content of cooler vent	kcal/kg clinker	67.2	72
		See Table 5.2.4 3b and 4	
Gross waste heat available	kcal/kg clinker	140	189
Heat required for drying		**Heat required for drying** **Case 1: raw material moisture 5%, coal 10%**	
Raw material	kcal/kg	113	113
Coal	Clinker	19.5	22.3
Total for drying		132.5	135.3
Net heat available for co generation	**kcal/kg clinker**	**7.5**	**53.7**
		Case 2: raw material moisture 3%, coal 10%	
Raw material	kcal/kg	67	67
Coal	Clinker	19.5	22
Total heat for drying	kcal/kg clinker	**86.5**	**89**
Net heat available for cogeneration	kcal/kg clinker	**53.5**	**100**

It would appear that if moisture in raw materials is 5% or more, sufficient heat is not available for cogeneration in the case of the six-stage preheater. If moisture in raw materials is 3%, then sufficient heat is available for cogeneration.

Quantity of cogenerated power would depend on the system installed (Rankine or organic Rankine or Kalina), according to its respective efficiency.

CHAPTER 3

Systems for Using Waste Heat for Cogeneration of Power

3.1 Systems for cogeneration of power in cement plants

Three systems are presently in use to cogenerate power in cement plants using waste heat from kiln exhaust gases and the cooler vent, depending on the grade of waste heat available.

3.1.1 Grades of waste heat available

Waste heat falls into one of the three grades shown below.

High grade:	Temperature: >600 °C		
	Uses	Preheat combustion air	
		Generate steam	
		Generate power	Waste heat recovery boiler
			Steam Rankine cycle (SRC)
Medium grade:	Temperature: 250-600 °C		
	Uses	Preheat combustion air	
		Boiler water preheat	
		Generate power	Waste heat recovery boiler
			Organic Rankine cycle (ORC)
Low grade:	Temperature: <250 °C		
	Uses	Domestic hot water	
		Generate power	Waste heat recovery boiler
			ORC

Designing Green Cement Plants
http://dx.doi.org/10.1016/B978-0-12-803420-0.00023-8

Copyright © 2016 BSP Books Pvt Ltd.
Published by Elsevier Inc.

3.2 Systems for using waste heat to generate power

1. Steam Rankine cycle

This is the most commonly used system in fossil-fuel-fired thermal power stations (TPS). Heat generated by burning coal (mostly), oil or gas is used to generate steam in a boiler. Superheated steam at high pressure is used to run a one- or multi-stage turbine coupled with the alternator.

Common components of this system are fuel firing system components like coal grinding and firing, combustion chamber, superheater, boiler drum/s, water preheater, economizer, electrostatic precipitator to clean flue gases, condenser to cool steam after the turbine, cooling towers to cool water and send it back into the system, and water feed pumps installed in appropriate locations.

When used to recover waste heat, the fuel preparation and firing systems are replaced by waste gases.

Boiler design is adopted to suit the condition of gases with regard to dust burden and the sticky nature of dust.

Other components remain the same, sized to suit the steam generated.

Fig. 5.3.1 shows the SRC adopted to recover waste heat.

2. Organic Rankine cycle

When incoming waste gases are at temperatures less than 300 °C, heat content is not enough to generate steam with the required degree of superheat. Thermal oil (like Mobiltherm 594) is used to receive heat from waste gases. This in turn transfers its heat to an organic fluid (like pentane) which is vaporized and is then passed through a turbine like the steam in an SRC.

Organic fluid is then cooled in an air-cooled condenser and returned to repeat the vaporizer cycle.

Cooled thermal oil is returned to the heat exchanger to be heated again by the waste gases and the cycle is repeated.

Plate 5.3.1 shows the Ormat ORC.

Fig. 5.3.2 shows the scheme of an ORC installed to generate power from the cooler vent in a cement plant.

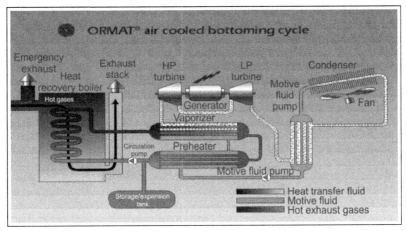

Plate 5.3.1 Organic Rankine cycle power plant. *(Source: Ormat International Inc.)*

In an ORC system, the turbine and piping system are smaller in size due to fluid density differences. Condensing pressure is generally above atmospheric pressure, eliminating the need for vacuum and gas purging equipment required in a steam condensing cycle.

Organic fluid has a much lower boiling point so that lower temperature heat sources can generate organic vapor. The organic turbine is a back pressure type and runs at much lower speeds and hence is more reliable and requires less maintenance.

Organic vapor becomes dryer as it expands in a turbine, unlike steam, which becomes wetter. Therefore it is not necessary to superheat organic fluid. Its freezing point is low. There is no freezing in the condenser even at very low ambient temperatures.

3. Kalina cycle (KC)

In this cycle a binary fluid containing ammonia and water is used as the medium that is evaporated by the waste gases and is passed through the turbine to run the alternator.

Binary fluid at any given pressure can boil or condense at a variable temperature.

Low-pressure vapor coming out of the turbine is first passed through a recuperator where it preheats the condensed ammonia water mixture returning to the boiler. It is then passed to a condenser. Water used to condense vapor is itself cooled in a condenser.

Fig. 5.3.3 shows a flow chart for the KC.

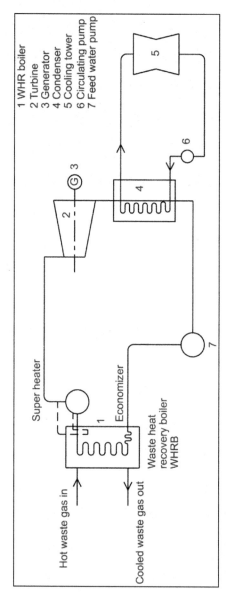

1 WHR boiler
2 Turbine
3 Generator
4 Condenser
5 Cooling tower
6 Circulating pump
7 Feed water pump

Super heater

Economizer

Waste heat
recovery boiler
WHRB

Hot waste gas in

Cooled waste gas out

Figure 5.3.1 Flow chart of SRC.

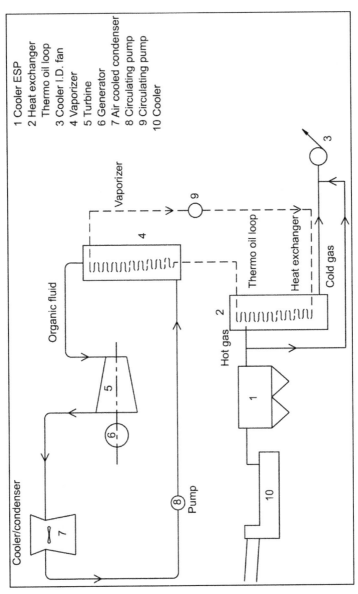

Figure 5.3.2 Flow chart of Organic Rankine cycle (ORC) shown on cooler vent.

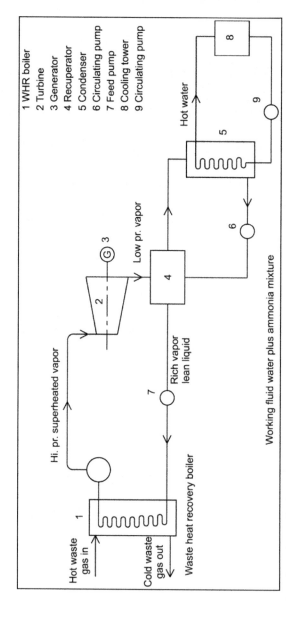

1 WHR boiler
2 Turbine
3 Generator
4 Recuperator
5 Condenser
6 Circulating pump
7 Feed pump
8 Cooling tower
9 Circulating pump

Hi. pr. superheated vapor

Hot waste gas in

Cold waste gas out

Waste heat recovery boiler

Rich vapor
lean liquid

Low pr. vapor

Hot water

Working fluid water plus ammonia mixture

Figure 5.3.3 Flow chart of Kalina waste heat recovery system. (*Source: article by NCCBM Scientists at 7th International Seminar.*)

In this system efficiency gains of up to 50% for low-temperature heat sources (200–280 °C) are claimed.

Conventional axial flow turbines can be used, as the molecular weights of ammonia and water are almost the same.

3.3 Salient features of these systems are summarized in Table 5.3.1.

Table 5.3.1 Salient Features of Various Waste Heat Recovery systems

Sr. No.	System	Salient Features	Advantages	Approx. Cost of Installation, Million US$ Per MW
1	Steam Rankine cycle	Waste heat used to generate steam from water which runs turbine to generate power	Suitable for waste heat sources with temps higher than 300 °C most commonly used system in thermal power stations reduces dust load in following Electrostatic Precipitators (ESPs);	1.3–1.5
2	Kalina cycle	Uses binary fluid ammonia and water as working fluid for heat transfer	Suitable for medium and low temp. waste heat sources	1.2–2.0
3	Organic Rankine cycle Ormat energy converter	Heat transfer from source to a thermal fluid in turn it transfers it to an organic fluid like butane or pentane	Uses air-cooled condenser highly reliable less maintenance	1.8–2.4

Source: Co-generation of Power Utilising Waste Heat in Cement Manufacture: Technological Perspectives, Presentation by NCCBM at 7th International Seminar

There are now a great many versions of these three basic systems with claims for improved efficiency and solutions for problems like handling dust.

The Thermowir system operating on the Rankine cycle incorporates a device for dividing gas flow into narrow steams and giving them rotary motion, which separates dust from the gas flow. Steam is produced within tubes employing heat in the flue gases.

Machinery manufacturers like Mitsubishi and Kawasaki, among others, have brought out their own systems.

3.4 In the following chapter some details of boiler design and other equipment used in WHR systems are described in brief.

3.5 Selection of system

From the previous section it is seen that the net heat available for cogeneration depends mainly on two factors

1. Heat content of preheater gases. This decreases with increased operational efficiency of a cement plant
2. Efficiency of the waste heat recovery itself. The ORC and KC are more efficient than the Rankine cycle. That is why they can produce with lesser net heat content.

Graph in Fig. 5.5.2 in Chapter 5 illustrates this.

CHAPTER 4

Machinery for WHR System

4.1. Typical configuration of a WHR system

Basically a WHR system is like the power generation system in a thermal power station (TPS) except that the source of heat for generating steam is readily available in the form of waste gases from the kiln and clinker cooler.

No investment is required to prepare fuel and fire it to produce hot gases.

This includes bringing in coal/oil, storing it, preparing it for firing like drying and grinding, metering it and feeding it through burners. This can be a sizable investment.

4.2. Waste heat recovery boiler (WHRB)

The main component as in the case of a TPS is the boiler with its parts, like the economizer and evaporating and superheating coils. Sufficient heating surface for transfer of heat must be provided to generate the desired degree of superheat and produce steam with a temperature above 300 °C and pressure of 15-20 bar.

Typically they are huge rectangular boxes through which gases flow and transfer heat to water and steam flowing in tubes. In the WHRB for pre-heater gases, gas flow is horizontal and tubes are vertical; in the WHRB for the cooler vent, gas flow is vertical and tubes are horizontal.

As mentioned earlier, the boiler is to be designed for gas conditions peculiar to cement plants:

1. Because of low-grade heat, boiler tubes should provide a large surface.
2. Preheater gases contain raw meal dust which can be sticky. Dust burden is also of the magnitude of ~70 g/nm^3.

To cope with these conditions most WHRB designs have plain tubes arranged vertically and gas flow is horizontal.

Designing Green Cement Plants
http://dx.doi.org/10.1016/B978-0-12-803420-0.00024-X

3. Pressure drop in a WHRB is on the order of 80-120 mm wg. System fans have to be designed to allow for this drop.

4. Gas velocities in the boiler should be such that dust does not tend to settle on tubes. Provision is made for periodic cleaning of tubes by soot blowers, sonic cleaners, mechanical rapping devices or steel shot dispersion, according to the character of the dust. The last two types are more common.

5. Tubes are subject to wear when dust is abrasive (like clinker dust).

6. A proper dust removal system should be integrated in the design of the boiler.

4.2.1. Sometimes supplementary firing is arranged to supplement heat in waste gases.

Another way to do this is to draw gases from the kiln riser duct at about 900 °C and mix them with preheater exhaust gases.

4.2.2. Water used in a WHRB should have a high level of purity and low hardness. Hence a water treatment plant/demineralizer has to be installed.

4.3. Condenser

Steam generated is taken to a back-pressure or condensing turbine in the case of a standard Rankine cycle. The condenser can be water cooled or air cooled, depending on the availability of water.

4.3.1. Dry condenser

These profit from humidity in the air as the enthalpy of ambient air increases with humidity. They are particularly suited for condensing organic vapor.

4.4. Water cooling towers

Wet cooling towers are capable of reaching temperatures below ambient through the evaporation effect of water in air.

4.5. Turbines

Turbines are back-pressure type. They can be multi-stage. In multi-stage turbines the high-pressure stage receives gases from steam generated by preheater gases and the low-pressure stage from the clinker cooler.

4.5.1 In the case of systems using the organic Rankine cycle, turbines run at slow speeds.

It is also possible to have a high-pressure topping turbine receiving vapor from the WHRB for preheater gases and a low-pressure bottoming turbine receiving vapor of organic fluid of the ORC that is vaporized by thermal fluid heated by the WHRB cooler vent. Both turbines are mounted on the same shaft and run the same generator.

See Flow chart 5.6.11 in Chapter 6.

4.6. The WHRB System must also have auxiliaries like a boiler feed pump, circulation pump, and water treatment plant, and also standard machinery and equipment and controls that generate electric power, like a generator, electrical switchgear and controls.

See Plates 5.4.1–5.4.3 for preheater and cooler WHRBs from two WHRB system designers and manufacturers.

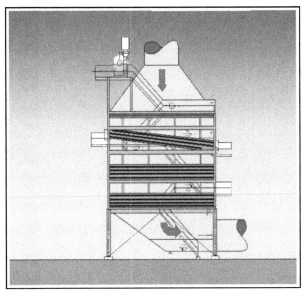

Plate 5.4.1 Waste heat boiler for clinker cooler. *(Source: Kawasaki plant systems waste heat recovery power generation).*

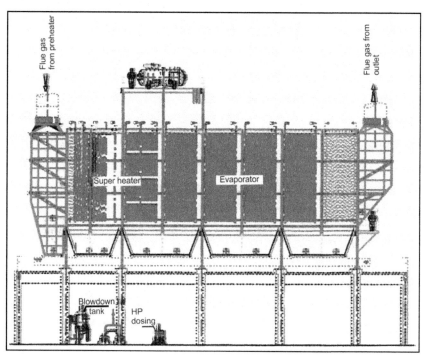

Plate 5.4.2 Waste heat boiler for preheater gases. *(Source: Thermax Brochure).*

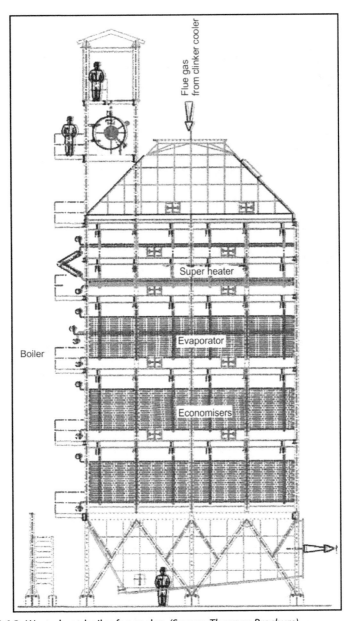

Plate 5.4.3 Waste heat boiler for cooler. *(Source: Thermax Brochure).*

CHAPTER 5

Potential for Generation of Power

5.1 All over the world, hundreds of waste recovery systems are now in operation, based on one of the three basic systems described. Based on operational experience, reasonably accurate estimates can be made regarding the amount of power that can be generated by way of WHR, given the specific operating conditions of a cement plant.

5.2 A yardstick has been devised for this purpose, that is, x kwh/t of clinker produced.

Each cement plant is different in some way or the other from another. Therefore this yardstick is to be used as a guideline for the detailed design of the WHR system project.

    5.2.1 Broadly speaking, it should be possible to produce by cogeneration power to the extent of

kwh/t Clinker	Stages of Preheater
35	4
30	5
25	6

5.3 Actual generation may be more or less dependent upon the seasonal requirements of drying to be met by waste heat gases. In Chapter 2 estimates of waste heat available in waste gases under different operational efficiencies of kilns and coolers were worked out. Details have been furnished to arrive at the net waste heat available.

5.4 Waste heat available in kiln gases and in cooler vents have also been shown separately so that individual or collective estimates of WHR systems can also be made.

5.5 The quantum of waste heat and degree of waste heat decides the type of waste heat recovery to be installed.

It is clear from that exercise that the quantum of waste heat generation decreases with the increase in operational efficiency of the cement plant.

Designing Green Cement Plants
http://dx.doi.org/10.1016/B978-0-12-803420-0.00025-1

5.6 A study of estimating potential for cogeneration carried out by NCCBM India in 2007 shows that the potential ranges from 3 to 5.5 MW in different plants depending upon availability of gases (after meeting drying requirements), actual temperatures of gases and operating efficiencies (number of stages of preheater).

Example:

> 3000-tpd cement plant producing OPC
> annual clinkering capacity ~1,000,000 t
> at 35 kwh/t, energy generated by WHR = 35 million units
> number of running hours per year = 330 × 24 = 7920
> therefore units per hour = 4419
> therefore power in MW = 4.4 MW

5.7 If the plant requires 80 units per ton of OPC, then its requirement of energy on hourly basis is 10,500 units. Considering various factors, the plant would opt for a maximum demand of 13 MW. Thus the cogenerated power is ~33% of the plant's normal requirements. After making allowances it can be safely said that the WHR power ranges between 20% and 30% of normal requirements.

5.8 Graph Fig. 5.5.1 shows the potential for power generation by installing a WHR system.

5.9 The graph in Fig. 5.5.2 shows the relation between power generation in kwh/t and the heat needed in kcal/kg clinker for different operating efficiencies. This will help with the selection of the WHR system to be installed.

Organic Rankine and Kalina cycles should be operating at efficiencies ranging from 45% to 55%, compared to the efficiency of ~37–40% of the steam Rankine cycle.

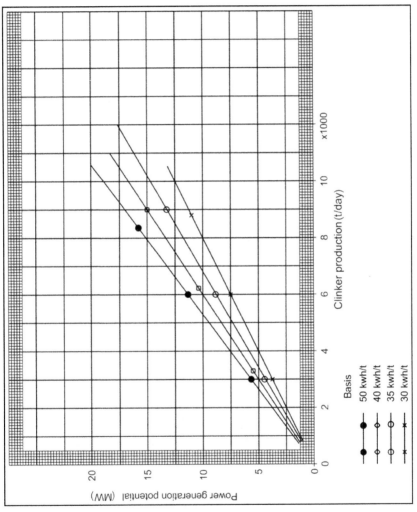

Figure 5.5.1 Graph showing potential for generation by WHR (waste heat recovery) system.

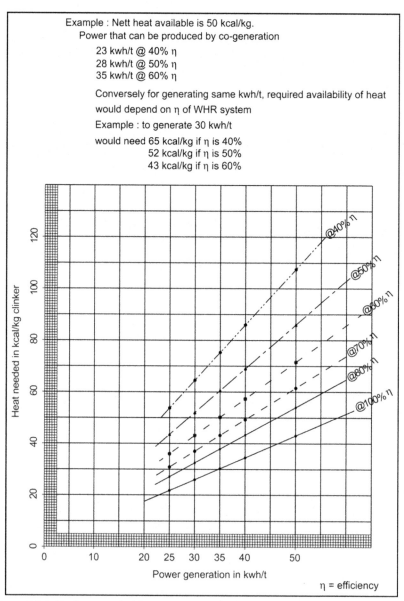

Example : Nett heat available is 50 kcal/kg.
Power that can be produced by co-generation

23 kwh/t @ 40% η
28 kwh/t @ 50% η
35 kwh/t @ 60% η

Conversely for generating same kwh/t, required availability of heat

would depend on η of WHR system

Example : to generate 30 kwh/t

would need 65 kcal/kg if η is 40%
52 kcal/kg if η is 50%
43 kcal/kg if η is 60%

η = efficiency

Figure 5.5.2 Net heat needed for cogeneration in kcalories/kg clinker for different WHR system efficiencies.

CHAPTER 6

Options for Location of WHR System and System Flow Charts

6.1 Apart from the choice of the system itself, that is, Rankine, Kalina or ORC, there are several ways in which the selected system can be incorporated into the kiln cooler system, depending on specific site conditions like the size of the plant, drying requirements and gas conditions.

> **6.1.1** Salient factors of these options are described below and are graphically presented in following flow charts.

6.2 Common or separate systems

These days preheater gases are used to dry raw materials as well as coal. Cooler gases are therefore generally available fully for cogeneration. In some cases, though, they may be used to dry fly ash/slag.

Therefore when a WHR system is installed both preheater and cooler gases are used for cogeneration.

The first option is therefore whether to have a common WHRS or separate systems for the two sources of waste heat.

The size of the plant is the deciding factor. Small plants prefer to have a common system.

Figs. 5.6.1–5.6.7 show different options for separate systems.

Fig. 5.6.8 shows a common system.

6.2.1 Separate systems for preheater gases

A WHRS can be located either before the preheater fan (a preheater fan designed to allow pressure drop in the WHRB) or after the preheater fan. See Figs. 5.6.1 and 5.6.2.

Full volume at maximum temperature goes to the WHRB and then to the mills. See Figs. 5.6.3 and 5.6.4.

Gases split to go to the raw mill and coal mill after WHR.

Designing Green Cement Plants
http://dx.doi.org/10.1016/B978-0-12-803420-0.00026-3

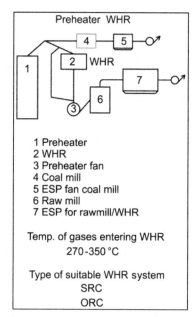

Figure 5.6.1 Flow chart for WHR system before preheater fan.

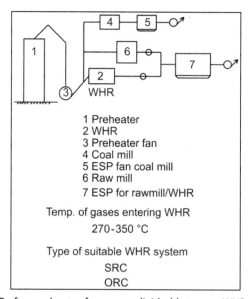

Figure 5.6.2 WHR after preheater fan; gases divided between WHR and mills.

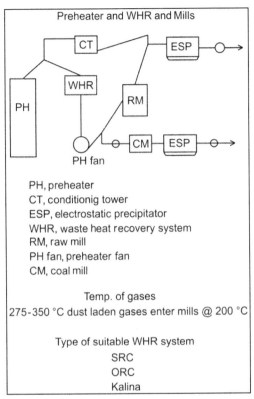

Figure 5.6.3 Flow chart for WHRS before preheater fan; gases split after WHRS to go to mills.

6.2.2 WHRB after the preheater fan; gases split three ways to go to WHR, raw mill and coal mill, all at exittemperature of preheater. A common ESP can be considered for WHR and raw mill. See Fig. 5.6.5.

In all cases, WHR would be installed in bypass mode so that the regular operation of the cement plant is not disturbed on account of WHR.

6.2.3 Four- to five- or five- to six-stage preheater

To augment exit gas temperature going to WHR, an arrangement can be made to tap part of the gases from one stage before the last stage. (This course is commonly adopted to increase the temperature of gases to be taken to mills for drying.) See Fig. 5.6.6.

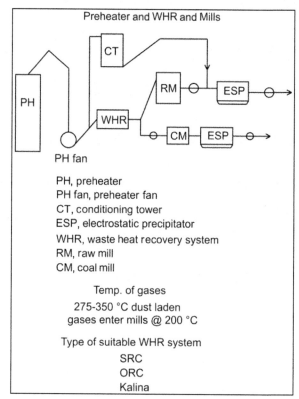

PH, preheater
PH fan, preheater fan
CT, conditioning tower
ESP, electrostatic precipitator
WHR, waste heat recovery system
RM, raw mill
CM, coal mill

Temp. of gases
275-350 °C dust laden
gases enter mills @ 200 °C

Type of suitable WHR system
SRC
ORC
Kalina

Figure 5.6.4 Flow chart for WHRS after preheater fan; gases go to mills after WHRS.

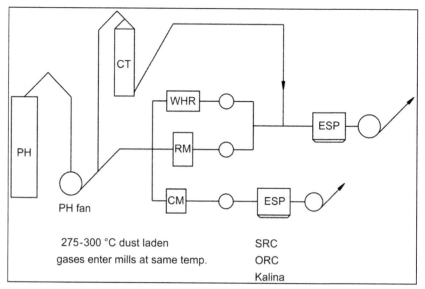

275-300 °C dust laden   SRC
gases enter mills at same temp.   ORC
                                         Kalina

Figure 5.6.5 Flow chart for WHRS after preheater fan, a variation of Fig. 5.6.2.

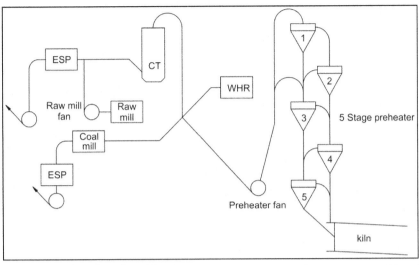

Figure 5.6.6 Flow chart for adding hot gases at two stages to increase temperature of hot gases from WHR.

6.2.4 One more option to increase gas temperature is to tap part of the gases from the kiln riser pipe and mix them with the preheater gases. This option is available when kiln gases do not need bypass. See Fig. 5.6.9.

6.2.5 Separate system for cooler vent air

One option is to tap the vent at the discharge of the cooler. Gases pass through ESP first so that only clean gases go to WHR.

For another option to augment the temperature of gases going to WHR, gases may be tapped at two points, once in the middle of cooler and once at the end.

There can be one fan but two are preferred. WHR is installed in bypass mode. See Fig. 5.6.7.

6.2.6 Common WHR for preheater gases and cooler vent air

A common WHR system with common facilities like the condenser and cooling tower is more appropriate for small plants. Gases are mixed before being taken to WHR. Location is selected so that losses due to pressure drop are minimum.

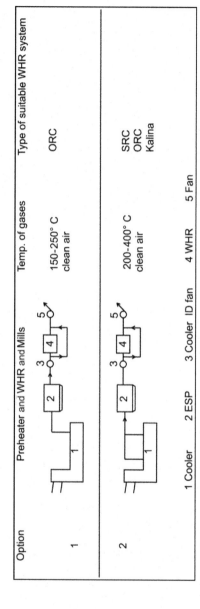

Option	Preheater and WHR and Mills	Temp. of gases	Type of suitable WHR system
1		150-250° C clean air	ORC
2		200-400° C clean air	SRC ORC Kalina

1 Cooler 2 ESP 3 Cooler ID fan 4 WHR 5 Fan

Figure 5.6.7 Flow chart for WHR system for cooler vent.

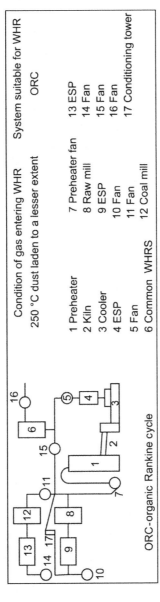

Condition of gas entering WHR   **System suitable for WHR**

250 °C dust laden to a lesser extent   ORC

1 Preheater	7 Preheater fan	13 ESP
2 Kiln	8 Raw mill	14 Fan
3 Cooler	9 ESP	15 Fan
4 ESP	10 Fan	16 Fan
5 Fan	11 Fan	17 Conditioning tower
6 Common WHRS	12 Coal mill	

ORC–organic Rankine cycle

Figure 5.6.8 Flow chart for a common WHR system for kiln and cooler.

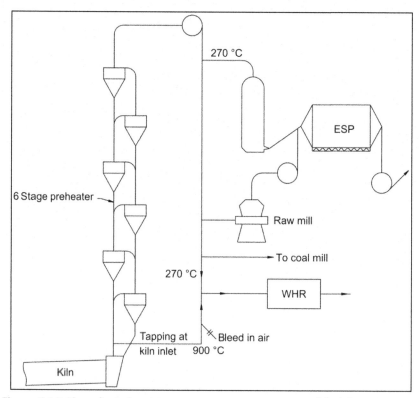

Figure 5.6.9 Flow chart showing arrangement to tap gases at kiln inlet to increase temperature of gases entering WHR.

In most cases gases to be taken to mills will be tapped before the WHR. WHR is in bypass mode as mentioned earlier. See Fig. 5.6.8.

6.2.7 This mode can also be adopted when hot waste gases are taken to an existing thermal power station, either to reduce fuel consumption or to increase capacity. See Fig. 5.6.10.

6.3 Turbine

In most cases where separate WHR systems are installed to recover heat from waste gases of the preheater and cooler, the turbine and following auxiliaries like condenser and cooling tower would be common. See Fig. 5.6.11.

It is also possible to install two turbines on the same shaft, one topping turbine receiving vapor generated by preheater gases and the other, a

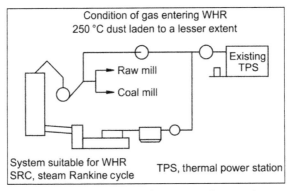

Figure 5.6.10 Flow chart for taking waste heat gases from preheater and cooler to existing TPS.

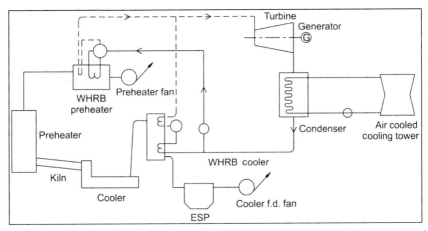

Figure 5.6.11 Flow chart for common turbine and separate WHRBs for preheater and clinker cooler.

bottoming turbine, receiving vapor generated by the cooler vent. See Fig. 5.6.12.

6.4 It is seen that a great many possibilities exist for installing a WHR system in a cement plant. They should be evaluated with the help of consultants and system designers before making a final decision.

The system selected should meet the criteria for maximum generation of power at minimum cost and should result in reduced GHG emissions.

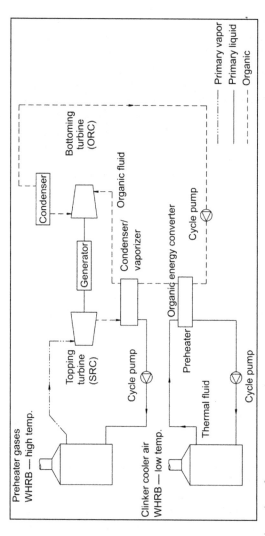

Figure 5.6.12 Flow chart for a WHR system showing two turbines on the same shaft. *(Source: Presentation by NCCBM at 7th International Seminar Co Generation of power using waste heat—Tecnological Perspectives)*

6.5 System flow charts

System flow charts are the best way of showing pictorially the principal features of a system, as evident from the various flowcharts referred to and attached. In the flow charts the types of WHRS most suited are shown.

Flow charts are also useful for compiling specifications for individual components of the system and inviting offers from vendors.

CHAPTER 7

Ordering WHR Systems

7.1 Whether a WHR system is to be installed in a new cement plant project or in an existing plant, the first step is to collect all relevant data on the cement plant, including site conditions, fuel to be used, fuel availability, power supply, and so on: any data that has a bearing on the selection and design of the WHRS.

7.2 It is best to engage experienced consultants to help and be associated with the project till its completion.

> **7.2.1** Presently designers like Ormat and Kalina, who invented the organic Rankine cycle and Kalina cycle, respectively, are themselves offering to design and supply the system directly or through their licenses.

> **7.2.2** Cement machinery manufacturers like FL Smidth and Kawasaki and electrical equipment manufacturers like Siemens are also offering to supply WHR systems (using the Kalina system). In such cases it is necessary to formulate the basic design (or actual) plant operating data and submit it to them.
>
> These companies would then either supply the complete system or equipment of their design and manufacture and furnish detailed specifications for auxiliaries.

> **7.2.3** Companies like ABB, Thermax, and Transparent Engineering Systems that are in business of designing and making waste heat boilers and turbo-alternators are also in a position to supply WHR systems for cement plants.
>
> Considering the special and uncommon nature of the product it may be best to opt for a turnkey or contract basis of supply.

7.3 Inquiry format

In this context reference is made to Section 9, Chapters 1 and 2 of the author's *Handbook for Designing Cement Plants*, in which preliminary steps for sending out inquiries have been explained.

Designing Green Cement Plants
http://dx.doi.org/10.1016/B978-0-12-803420-0.00027-5

7.3.1 An example format for inviting offers for WHR systems is shown as Annexure 1.

7.4 Technical specifications and performance norms

Inquires sent out should be either for a specific system, like SRC, Kalina or ORC, on the advice of consultants, or vendors may be asked to select the best to suit specific operating conditions, providing reasons.

7.4.1 Performance norms should be clearly spelled out, against which actual operation can be measured. Certain aspects of performance should be guaranteed against penalties.

7.4.2 Vendors must also furnish detailed technical specifications following prevailing norms for different machines, such as heat exchangers, boilers, condensers, cooling towers, turbines, generators, and so on.

7.5 How to measure efficiency of operation should be specified, as should an agreed protocol for proving it.

7.6 Considering the various options available and the variations in operating conditions, it is best to engage experienced consultants who can recommend how to meet requirements in an optimal way.

DEOLALKAR CONSULTANTS

CEMENT TECHNOLOGISTS & DESIGNERS OF CEMENT PLANTS

B-32, Shanti Shikhara Apartments, Raj Bhavan Road, Somajiguda, Hyderabad-500082

Phone: 040 23317803/66628267

E-mail: spdeolalkar@gmail.com

ANNEXURE 1

Enquiry for a waste heat recovery system for a 10,000-tpd capacity cement plant

1. Basic cement plant data

Size and capacity:	10,000-tpd dry process kiln with six-stage preheater and calciner and air-quenching cooler
Operational parameters:	Specific fuel consumption: 700 kcal/kg clinker
	Specific power consumption: 80 kwh/ton cement
Site conditions:	Sea level
	Ambient temp. 30 °C
Raw materials and coal:	Raw materials: 1.55 t/t clinker

		Moisture: avg.: 4%; max: 6%
	Coal:	avg. useful calorific value: 4500 kcal/kg
		Moisture avg.: 10%; max 12%

Power supply:	Incoming from grid: 132 kv 3 phase, 50 cycles
	HT (high tension) within the plant: 11 kv 3 phase, 50 cycles
Capacity of raw mill:	800 tph
Capacity of coal mill:	90 tph

Availability of water: fair from perennial river and ground water. Cost of electricity purchased from grid: Rs 4.0 per kwh including maximum demand charges.

2. Inquiry for firm quotation for a complete waste heat recovery unit to produce electricity from waste gases from both preheater and clinker cooler.

3. You may consider the best option from among

separate units for preheater and cooler or a common unit,
common Rankine cycle (generating steam from water), and/or

Ormat Rankine cycle or Kalina Cycle.

Taking into account requirements for drying as spelled out above.

4. Your offer should include

1. **Plant and machinery** to be designed and supplied by you
   - With broad specifications as also bought outs
   - Budgetary price for civil construction
   - Electrical switch gear and controls
   - If steam Rankine cycle, provision required by cement plant for water
   - Flow chart of the system

   Also, layouts showing the plan and elevations of a similar installation fitted into the kiln and cooler section of the plant.

   If Ormat cycle arrangements are required for storing and thermal oil and organic fluid (are these to be imported) and for feeding to the system

   - Plant and machinery required, with price in $

2. **Performance parameters** like

   - Power produced in MW at 11 KV 3 phase, 50 cycles
   - Power produced in kwh/t clinker and as percentage of requirement of grid power
   - Efficiency of the plant
   - Running hours per year
   - Cost of electricity produced in the plant
   - Power consumed in the plant and net savings
   - Annual savings in Rs/$
   - Total capital costs in Rs/$/MW
   - Payback period in years

3. **Commercial terms**

   - Prices FOB, FOR as the case may be in Rs/$
   - Terms of payment
   - Guaranteed delivery
   - Performance guaranteed against penalties and terms thereof
   - Warranties for design and workmanship

CHAPTER 8

Capital Costs, Savings, and Payback Period

8.1. Feasibility of generating power from waste heat will be evaluated against the investment required to install a WHR system, savings that would accrue and the consequent payback period, which is often considered a yardstick measure of investment effectiveness.

> **8.1.1.** WHR is an attractive proposition from other angles, such as
>
> > **1.** savings in fossil fuel
> > **2.** reduction in greenhouse gases
> > **3.** reduced cost of power generated by WHR system
>
> However the tangible gains are measured in terms of savings versus investment.

8.2. Even for the same size of the clinkering unit, the type and size and hence costs of the WHR system to be installed will vary from plant to plant. Hence overall investment in millions of dollars has no great significance. Investment in a cement plant is universally measured by the yardstick of Rs/ton of annual capacity.

> Similarly, in the case of WHR systems, the yardstick of million $/MW is a commonly accepted.

8.3. The investment expressed by this yardstick varies in a small range for the same type of system and differs for different types of systems.

> **8.3.1.** Data gathered from the report presented by NCCBM (India) in the year 2000 at the 7th International Seminar on various WHRs quotes indicative investment costs in million $/MW

Designing Green Cement Plants
http://dx.doi.org/10.1016/B978-0-12-803420-0.00028-7

As follows:

Type of System	Investment in Million $/MW
Standard Rankine cycle	1.3-1.5
Kalina cycle	1.2-2.0
Ormat Rankine cycle	1.8-2.4
Thermowir variation of Rankine cycle	1.1-1.4
Mitsubishi and Kawasaki variations of Rankine cycle	1.3-1.5

8.4. Payback period ranges between 2 and 4 years, depending on the cost of grid power. Higher cost of grid power, greater the savings and hence shorter payback period.

Not much data are available to be more accurate in this respect.

8.5. There are a great many operating WHR systems generating electricity in cement plants. The majority are using the standard Rankine cycle. A good many systems using the Ormat Rankine cycle and a few using the Kalina cycle are also in operation.

8.6. It is more or less established that the power generated by WHR ranges from 30 to 45 kwh/ton clinker, with a median value of 35 kwh/ton clinker.

8.7. At 30 kwh/ton, a 10,000-tpd kiln has the potential to generate ~12.5 MW of power. At 1.5 million $/MW, the investment required to install a WHR system is 18.75 million dollars.

Capital costs for a cement plant are ~75 $/ton of annual capacity.

Therefore for a 10,000-tpd or 3.5-mtpa cement plant producing OPC, capital costs would be ~$265 million.

Investment costs of 18.75 million dollars for a WHR system are thus only about 7%.

8.7.1. Power generated from a WHR system costs much less than grid power and also less than the cost of power from a captive power plant. It is estimated to be Rs 2.5/kwh (~4.5 US cents) as compared to the cost of grid power at Rs 5 (9 US cents).

Payback period may therefore be reasonably short.

Hence, generally speaking, investment in a WHR system is an attractive and economically viable proposition for new projects and also for existing projects.

CHAPTER 9

Design and Operational Aspects

9.1. Previous chapters have described the various aspects of design and engineering of WHR Systems and integrating them in new and existing cement plants.

    9.1.1. Generating power is not the only use for waste heat from kiln and cooler gases. There are other possibilities, such as:

        1. to heat water

        2. to produce low-pressure steam for preheating and indirect drying

        3. for air conditioning and making ice products

        4. for demineralization of water.

        These can also be considered at the time of initial planning

9.2. In countries like India where power shortage is substantial and chronic, cement plants install captive thermal power stations sized to supply almost 40% of their power requirement. Fuel used therein is mostly coal (fossil fuel) which contributes to greenhouse gas emissions. It has been seen that WHR ystems can generate up to 30% of power requirements for a cement plant.

In the case of new projects it should be possible to integrate the WHR system with the proposed captive power plant.

    9.2.1. In the case of an existing cement plant and a TPS, it should be possible to take waste gases to the existing TPS and augment its capacity rather than put up a separate WHR system.

    Auxiliaries can be sized from the beginning to produce an enhanced quantity of power. Alternatively a module can be added to the existing TPS.

9.3. Special circumstances for running WHR systems in cement plants

    1. Firstly, operation of the cement plant is the main activity always the first priority. WHR Systems are therefore always installed in

Designing Green Cement Plants
http://dx.doi.org/10.1016/B978-0-12-803420-0.00029-9

bypass mode. Further, turbines must run at > 75% of their capacity. Therefore, for the WHR System to be operative, the cement plant must run at correspondingly high capacity.

High-efficiency cement plants of the present day need to be operated close to their design capacities.

2. The power plant must be reliable and must run for about 90% of the time available. Cement kilns are designed to run for 330 days of the year. The WHRB is operative when the kiln is operative. It is idle when the kiln is shut down.

3. Waste gas will first be used to dry raw materials and coal. Mills are designed to run for 20 hours in a day, whereas the kiln runs for 24 hours. Thus for about 4 hours every day the availability of waste heat to generate heat is substantially higher. How best to cope with this periodic repetitive situation needs to be evaluated carefully at the planning stage.

4. Whereas the temperature of gases entering the raw mill can be as low as 200 °C, because of the large volumes needed in vertical mills it may have to be as high as 300 °C in the case of coal mills. How to obtain such a differential in temperature from one source needs to be evaluated.

5. Problems related to dust and its sticky nature in preheater gases have been described. Clinker dust is coarse and abrasive. These special conditions have to be taken into account when designing WHRBs.

6. Kilns sometimes go into "upset" mode, which may last for two or more hours. In this period, the volume and temperature of the cooler vent is vastly different (gas volumes higher by 30% and temperatures of more than 300 °C). The WHR system has to cope with these abnormal conditions now and then.

7. In the case of a WHRB on a cooler vent, in most cases the Electro static precipitator (ESP) is installed before the WHRB. If a dust collector is still required after the WHRB, it is much smaller in capacity

9.4. A WHR system should therefore be selected by carefully taking into account various pros and cons with the help of renowned consultants in the field.

9.5. Layouts with the WHRB system

In new plants, it should be comparatively easy to incorporate a WHRB system with its auxiliaries.

If the system is based on the Rankine cycle, generating steam to run the turbine and, if the plant already has it, a captive TPS, auxiliaries like the condenser and cooling towers can be located alongside the existing facilities.

9.5.1. If WHR is of the Kalina or ORC type, separate arrangements have to be made for installing corresponding vaporizers and condensers. Storage facilities for thermal oil and organic fluid have to be created.

In plants located in areas where water is scarce, air–cooled condensers and cooling towers are installed.

Fig. 5.9.1 shows the WHR system layout for the preheater and cooler as well as the main thermal power station.

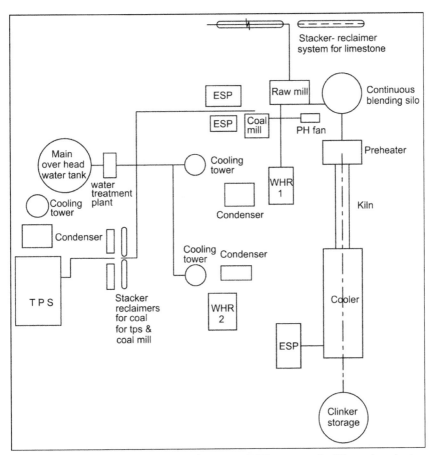

Figure 5.9.1 Layout showing fitting of separate WHR systems for kiln and cooler in a cement plant layout. *(Existing Thermal Power Station - TPS -& its auxiliaries have also been shown).*

9.6. Availability of Technology

WHR Systems are now widely installed in cement plants and are designed to meet specific requirements. Availability of various technologies should not be a problem even in countries that are just beginning to introduce WHR systems.

Cement machinery manufacturers like F.L. Smidth and Kawasaki are also offering WHR systems through their international networks. China is well ahead in this field.

The import content of some of the technologies like ORC is not clear. If it is substantial, it may be a deterrent in some countries.

9.7. In the case of new projects, funds should not be a problem, as the WHR system costs only about 5–7% extra. Small and older plants which actually have greater potential to generate power by WHRS may have some difficulty raising funds. However, in these cases payback periods are shorter.

CHAPTER 10

Evaluation of Results of Installing WHR Systems

10.1 Success of the installation and operation of the WHR System should be measured against objectives that prompted its installation in the first place.

10.2 There are three main objectives.

1. reduced costs of operation
2. savings in consumption of fossil fuel
3. reduction in emissions of greenhouse gases.

10.3 Coal saved and reduced costs of operation

Costs of operation are reduced because of generation of power at a reduced cost.

In generation of power by WHR, heat is supplied by waste gases. No additional fuel is required.

Example:

> cement plant capacity: 3000 tpd/1 mtpa opc
>
> power generated by WHR: 4 MW at 32 kwh/t
>
> heat input required per hour assuming 40% efficiency = 9 million kcal/h
>
> assuming average calorific value of coal = 4000 kcal/kg,
>
> coal consumption to produce this heat = 2.25 tph

Therefore coal saved per year (330 days × 24 h) = 17,800 t.

10.4 Savings in energy costs

Assume energy costs of $0.09/kwh and specific power consumption of 80 kwh/t

Designing Green Cement Plants
http://dx.doi.org/10.1016/B978-0-12-803420-0.00030-5

Energy cost in cost of production $= \$7.2/t$ cement

After installing WHR, energy distribution of 80 kwh/t would be:

from WHR (at ~80% of installed capacity)	30 kwh
from grid	50 kwh

Grid power costs of $.09/kwh

Let cost of fuel in cost of grid power be 40% (for illustrative purposes only)

Cost of WHR power would be $0.045/unit

Total energy costs would be $ 0.07/unit

a savings of ~22%

On 1-million-ton production, savings would be $1.6 million per year.

10.5 Savings in emission of greenhouse gases

Emission of CO_2 without WHR is 0.80 t/t of cement (OPC) divided as 0.51 t due to calcination and 0.29 t due to combustion of fuel.

With WHR, emissions due to combustion correspond to 30% extra production. Therefore emissions per ton would now be 0.29/1.3 $=0.22$ t/t.

Total emissions would be $0.51 + 0.22 = 0.73$ t/t

a reduction of ~8.75%

Annual reduction in emissions of CO_2 can be worked out as follows:

production in one year $= 1,000,000$ t clinker

CO_2 emitted without WHR $= 800,000$ t

CO_2 emitted with WHR $= 730,000$ t

saving $= 70,000$ ton per year.

10.6. It is seen that on all three counts mentioned in paragraph **10.2**, a WHR system has a positive and beneficial impact not only on the plant but on the industry and on the earth as a whole.

10.6.1. Further, for new plants, the extra expenditure in capital costs to install WHR is small, less than 10%, and payback periods are reasonably short, at between two and four years.

10.6.2. Systems already in operation have proved to be fairly reliable, with about 90% availability.

10.7. The benefits can increase further if blended cements are made and the plant uses alternative fuels. It is thus a **win-win-win** situation.

RECOMMENDED READING

1 Co generation of Power Utilizing Waste Heat in Cement Manufacture-Technological Perspectives 7th NCCBM International Seminar 2000

2 Generation of Power thru Waste Heat Recovery in Cement Plants-Overview and Prospects by Pradeep Kumar NCCBM

3 Co-generation – success story in Indian Cement Industry by R A Sharma Master Consultancy & Productivity Pvt Ltd

4 Energy Availability & Recovery for Dry Cement kiln Systems Case Study by Tashen Engin & Vedat Ari Science Direct – Energy Conversion & Management 2004

5 Recovering Heat for Energy by Dr. Thomas Burki et al, ABB Switzarland

6 Waste Heat Recovery – Cement Industry by K. Umamaheshwar Gulf Coast Clean Energy Application Center 2011

7 Energy Balance & Co-generation for a cement plant by Shaheen Kurara, R Banerjee & U Gaitonde Applied Thermal Engineering 2001 Pergamon

8 Organic Rankine Cycle (ORC) Power Plant- Waste Heat Recovery by L Y Bronicki – Ormat International Inc

9 Recovery of Industrial Heat in Cement Industry by ORC Process by Hilel Legmann Ormat Technologies Inc.2011

10 Future Trend – WHR in Cement Plants by Sameer Bharadwaj KHD Humboldt Wedag- Presentation at Green Cemtech 2010

11 WHR Power generation for Cement Plant Presentation at Energy Saving Seminar – Mexico by Kawasaki Plant Systems Ltd 2007

12 Ningguo Cement Plant 9.1 MW WHR & Utilization for Power Generation Project of Anhui Conch Cement Co. Ltd Monitoring Report 2–2010

Other Aspects of Green Cement Plants

Contents

List of Figures

List of Plates

List of Tables & Annexures

CHAPTER 1

Introduction

1.1. So far we have dealt with aspects that are integral to the process of making cement. However there are many other aspects which also must be considered when designing green cement plants.

1.2. Green buildings

There is great emphasis now on designing green buildings. The meaning of the green is the same: to conserve natural resources, to conserve energy, and to use recycled and locally available materials with reference to the design and construction of buildings.

This concept can be profitably used in designing office buildings, laboratories, and residential premises for cement plants.

1.3. Water conservation

Water is getting scarcer every year although it does get replenished by rains. Conservation of water is assuming great importance all over the world, in all walks of human activity. It can take the shape of rain water harvesting or treating and recycling gray water, or even creating reservoirs in exhausted quarries.

1.4. Landscaping and green belts

Most cement plants own mines from which limestone is extracted and used to supply the cement plant. Over a period of time, the removed overburden and contamination in stone that is to be thrown away, assume ugly proportions.

Trees are cut down and deforestation occurs in the process of obtaining limestone.

Landscaping attempts to beautify the vicinity of mines and cement plant and create green belts that act as dust and sound barriers.

1.5. Electrical energy

Cement plants run on electrical energy. There are literally thousands of electrical appliances and instruments, domestic and industrial, in cement plants. Presently energy efficiency bureaus in all countries are acting as watchdogs of efficient electrical energy use by setting standards of performance for various electrical devices, beginning with the common light bulb.

Adoption of energy efficient devices and appliances certainly reduces energy consumption and also helps reduce greenhouse gas emissions.

1.6. Renewable energy sources

There is great emphasis on saving fossil fuels in the process of making cement and also on waste heat recovery. The principal source of electricity is still thermal power.

The logic of conservation of natural resources leads one to seek sources of energy that are renewable, such as wind, solar, and hydraulic power. A great many cement plants are making use of wind power and solar power in a significant way all over the world. China and the US are leading the pack.

Therefore a green cement plant must take a close look at the possibility of using these alternative sources of energy and reducing dependence on fossil fuel power.

A new word has already been coined: **ecotricity.**

1.7. Mining

As mentioned earlier, open cast mining adapted to obtain limestone leaves a lot of debris around the actual operating faces. As mines develop, mining goes deeper and wider to cover the area containing usable limestone.

The operation denudes areas of trees and shrubs. Overburden that needs to be thrown away assumes ugly proportions.

The entire area may be called the **footprint** of the mining operations. In designing green cement plants, an attempt is made to make this footprint as small as possible by using more scientific mining methods and by planning in advance how to dispose of the overburden and rejects from the mines.

A conscious effort is made to replace uprooted trees by planting new ones and to preserve the ecosystem.

1.8. The above activities may not have a direct bearing on the actual operation and design of a cement plant. Some activities like designing and installing alternative power from renewable sources are often entirely independent of the cement plant and are installed at places far removed from it. For example a wind farm is used at a place most suited for it.

Yet taken as a whole, this effort is a part of the green cement plant conception.

Therefore in evaluating the project as a whole with regard to total investment and returns all such steps should be taken into account. Their total impact on sustainable development should be evaluated.

1.9. These aspects are dealt with sequentially in the following chapters. It is not possible to be precise regarding capital investments and details of design in all cases. The purpose of the book is served if their importance is highlighted and if in the planning stages of a new cement project a conscious effort is made to include as many of them as possible in the design and execution of the project.

The cement plants of the future will then be truly **green**.

CHAPTER 2

Green Buildings

2.1. Green buildings

As mentioned in Chapter 1, green buildings are constructed with a view to conserving natural resources and making maximum use of natural light and ventilation so that dependence on artificial lighting and air conditioning/heating is minimal.

2.2. Factory buildings

Factory buildings housing machinery need to be designed to suit the machinery to be installed and to suit the flow of materials in and out. Lighting is arranged to suit the needs of the operations being carried out and to suit measurements taken. The type of lighting selected can make an impact For example, ordinary tube lights can be replaced by CFL tubes.

2.3. Ventilation

Ventilation should be arranged to make occupation of the buildings comfortable for the operators. The building may house dust collectors to arrest dusts arising from operation or flow of materials. The cleaned air/gas has to be vented out of the building. Cross ventilation generally cools the buildings and should be used to the maximum extent.

2.4. Office buildings and laboratories

Besides main plant buildings a cement plant contains lots of other buildings, such as stores, go downs, laboratories, workshops, and so on. They have their own functions and are designed to discharge that function efficiently and economically.

2.4.1. Housing

Most cement plants have to provide housing facilities at least for core workers, supervisors, and executives. They are generally located close to the cement plant itself and take the form of a colony or miniature township. There is plenty of scope here

Designing Green Cement Plants
http://dx.doi.org/10.1016/B978-0-12-803420-0.00032-9
215

to design housing to suit local climatic and environmental conditions that keep the green principle in mind.

2.5. Norms for green buildings of various types

To conform to the category of a green building, there should be first a set of standards that qualify the building to be rated as green. Fortunately such standards are available. There are a number of countries who have created such standards.

2.5.1. Rating system

IGBC[1] has designed a rating system to judge the "greenness" of a building. In designing the factory and other buildings it is a good idea to refer to it to make it easier to obtain the green building rating upon completion.

CII (Confederation of Indian Industries) has relevant publications on the subject. A list of CII Publications is attached.

2.5.2. Aspects covered by the norms

It is seen from perusal of the CII Norms that they are an excellent guide for designing all types of buildings commonly found in cement plants. Over and above that the rating system helps the designer to know the degree of greenness achieved as construction progresses.

2.5.3. Guiding norms in brief

In Annexure 1, guiding norms have been given in brief to serve as ready reference. However the designers would do best to refer to the original norms.

List of CII publications on green buildings

1. "LEED" 2011 for India for New Construction & Core & Shell. Detailed Reference Guide
2. Landscape Directory
3. Directory on Green Building Materials & Service Providers
4. IGBC Green Factory Building Rating System: Pilot Version
5. IGBC Green Homes Rating System Version 1.0

---

[1] India Green Business Centre.

ANNEXURE 1

Extracts of norms for green buildings for factories & homes

The following extracts of norms have been culled from the IGBC Publications

1. IGBC Green Factory Buildings Rating System: Pilot Version
2. IGBC Green Homes Rating System Version 1.0

They are a good guide for various aspects of green buildings, to be kept in mind while designing factory buildings and homes connected with cement plants.

The rating system is voluntary. It is based on currently available materials and technologies. The objective is to develop eco-friendly factory buildings and homes that are energy and water efficient and healthy and productive.

1. **Selection of site**

   Building to comply with local statutory and regulatory options control soil erosion—National Building Code of India.

2. **Accessibility to public transportation**

   This is important due to air pollution from personal automobiles.

3. **Basic amenities within the building** should include a first aid center, nursery, lockers and showers, canteen, guest house, and cyber café/ Internet facility (one for every ten people).

4. **Topography and landscape**

   Site disturbance should be avoided by retaining natural topography.

5. **Heat island effect**

   Thermal gradient between developed and undeveloped areas cover 50% of roof by vegetation—75% area open parking by trees—roof over parking or parking in basement. Heat island means – an area of temperature higher than that of surrounding because heat is prevented from dissipating naturally to areas of lower temperatures.

6. **Differently abled people**

   The building should be user friendly to differently abled people.

7. **Water conservation**

   Mandatory 50% of roof and non-roof runoff should be under RWH. Increase ground water table or reduce water consumption by effective rainwater management. Rainwater storage for two to three days of rainfall.

Plant drought-tolerant trees/shrubs.

Non-process water treatment either individual or common effluent treatment plant (ETP).

ETP-treated water should meet the standards of quality laid down by CPCB.

8. **Energy conservation**

   Avoid use of CFC-based heating, ventilating, and air conditioning (HVAC) equipment

   Avoid use of HCFC-based refrigerants

   Optimize energy performance

   Prescribe minimum energy performance

   Provide metering to monitor energy consumption

9. **On- and off-site promotion of renewable energy**

   Minimum of 5% energy used in building, excluding process energy from renewable green power source. Promote investment in renewable energy 50–100% off site.

   Eco-friendly generation of captive power using low emission biofuels.

   Use diesel generator (DG) sets certified by CPCB and ISI-rated generators.

10. **Eco-friendly captive power generation**

    Low-CO_2-emitting fuels to be used in CPP-biofuels.

    Use DG sets certified by CPCB and ISI-rated generators in both cases.

11. **Material conservation**

    To ensure effective management of non-process-generated wastes and arrange for:

    Their segregation (organic, plastic, glass, e-wastes, etc.)
    Waste reduction during construction so minimum sent to landfill
    Use materials with recycled content
    Use forest department certified wood

12. Indoor environment and occupational health

Protect nonsmokers from passive smoking

Minimum fresh air requirements for air-conditioned buildings.

Install fresh air delivery system to ensure minimal air changes for buildings with forced ventilation.

Avoid use of asbestos.

Day lighting.

Provide good day lighting: daylight factor minimum of 2% for minimum of 50% of total floor area. Daylight factor has been defined in the guidelines.

Use paints with low VOC.

Use eco-friendly housecleaning materials.

Special points for green homes

Most of the guidelines listed above are also applicable to green homes, with suitable modifications. A few are applicable specifically to homes.

1. Landscape

Limit landscapes that consume large quantities of water. Turf area to be <20–40% of total plot area.

Plant drought-tolerant species of trees.

At least 50% of landscape planting beds must have drip irrigation.

Survey water table in area.

Factors to be considered are weathering, fractures, and joints for rocky sites and thickness of aquifer in sedimentary rocks.

2. Gray water treatment

Gray water is waste water resulting from bathrooms, kitchen washing, and so on. Provide facility for treating at least 50% of gray water for gardening supply, etc.

Reduce demand for fresh water by using gray water for landscaping.

3. Energy-efficient lighting and electric appliances

install energy meters wherever possible for monitoring consumption

refrigerators: minimum 3 star rating by BEE

install solar water heating system

lighting luminaries: minimum 3 star BEE rating

power density ~20% less than baseline values

select efficient motors for pumps, lifts, etc.

4. Organic waste segregation

Install on-site waste treatment technology to prevent such waste being sent to landfills.

5. Indoor environment and occupational health (same as for factories exhaust system)

Kitchen and bathrooms to be better ventilated.

These are extracts only. Please refer to the original books referred to above.

CHAPTER 3

Water Conservation and Rainwater Harvesting

3.1. Water conservation

Cement plants have been conserving water in their plants from the beginning as most cement plants have had to make their own arrangements to obtain water required for the plant and for drinking and household purposes.

3.1.1. Cement plants procure water from the nearest perennial sources of water like rivers and streams by digging wells in their beds and pumping it and storing it in the plant/quarries/housing colony.

3.1.2. Plenty of water is required even in dry process cement plants to cool bearings, compressors, after-coolers, gearboxes and for conditioning towers preceding ESPs. All water used for cooling is invariably collected and taken to a cooling pond and recirculated in the system. Only 10-15% water is added to allow for loss by evaporation.

3.1.3. Process water is not required in a dry process cement plant. However if an ESP is used to clean preheater exhaust gases, a cooling tower is necessarily installed to bring down the temperature to about 140 °C. Gases are cooled by spraying water on the gases in the cooling tower. Water evaporates and is consequently lost. This is a significant quantity.

3.1.4. This loss of water can be avoided if the ESP is replaced by a bag filter. However penalty there a penalty for the higher pressure drop in the bag filter and the necessity of cooling gases to ~120–140 °C by admitting ambient air to suit the materials of bags. If glass bags which can stand a temperature of ~275 °C are used this dilution can be avoided. Generally speaking the ESP can be avoided at the design stage itself if the plant is located in an area of scanty rains and water scarcity.

Designing Green Cement Plants
http://dx.doi.org/10.1016/B978-0-12-803420-0.00033-0

Performance of the ESP is uncertain during startup and closing down periods. Presently the trend is to avoid an ESP for this reason also.

3.2. Thermal power station/waste heat recovery

When a cement plant installs a thermal power station (TPS) to supplement or to ensure availability of power on a continuous basis, it requires large quantities of water for generation of steam.

The steam is condensed in condensers and returned to the circuit. Water used to condense steam is itself cooled in cooling towers operating in a closed circuit; that water is used again by recirculation. Therefore, only makeup water is required. The same is true of waste heat recovery boilers. Even where DG sets are used to generate power, diesel engines are cooled by water which in turn is cooled in cooling towers and returned to the circuit.

Air-cooled cooling towers and condensers are now used to minimize use of water.

3.3. Water for laboratories, offices, and domestic use

This water is generally wasted after use, though sewage water can be used after treatment for nondrinking purposes like gardening. As a matter of fact authorities who sanction a cement plant project stipulate that a sewage treatment plant has to be installed in the plant. There should be **zero effluent discharge** from the plant.

3.4. Mines as a source of water

Often, as mines get developed, underground resources of water become available and actually supplement the main source of water. Pits in excavated/exhausted mines can be used to serve as reservoirs of water. These are available year round for mining machinery and crushing plant when located in mines.

The above-mentioned authorities also stipulate that planning of mines includes creation of reservoirs which can recharge groundwater and increase the height of the water table.

Often, currently, used mines are consciously developed and landscaped to serve as recreation or picnic spots. Reservoirs in mines thus serve a dual purpose, as a source of water and as lakes. When treated the water can also be used in swimming pools.

Presently there is great emphasis on greening of the plant and its surroundings, including the housing colony. Green belts are created around the plant and colony to serve as dust and sound barriers. It is mandatory to create such belts between the plant and the highway/township.

Water in mines can thus be put to many uses, thereby saving water from the main and perennial source of water.

3.5. Rainwater harvesting

In the context of cement plants, rainwater harvesting (RWH) has many dimensions.

1. Rainwater is collected and stored in natural/artificial ponds or lakes to counter the salination of groundwater in coastal areas.

For this purpose check dams are constructed across streams and rivulets.

2. A system called "garland canals" is constructed to collect the groundwater and lead it to reservoirs in quarries or reservoirs created by check dams.

This water can be used in the cement plant for manufacturing, in captive power plants, and for domestic use in colony.

As a matter of fact many cement companies are supplying water for drinking purposes and for agricultural purposes to neighboring communities on an increasing scale. Some have installed desalination plants also.

The authorities sometimes stipulate that the cement company should not draw water from an adjoining river/stream.

3. RWH is used to recharge bore wells within the plant's own area and colony.

Water is collected from rooftops and led through pipes to collection pits near the bore wells to recharge them. Water is then available year round, even in summer months.

3.6. Extracts from a typical letter of consent for a cement plant project at a green field site or for an expansion show the emphasis the authorities are putting on water conservation. Cement plants of the future will have to be green. See Annexure 1.

3.7. A typical check dam and system of garland canals is shown in Fig. 6.3.1. A typical system to collect rainwater from roofs is shown in Fig. 6.3.2.

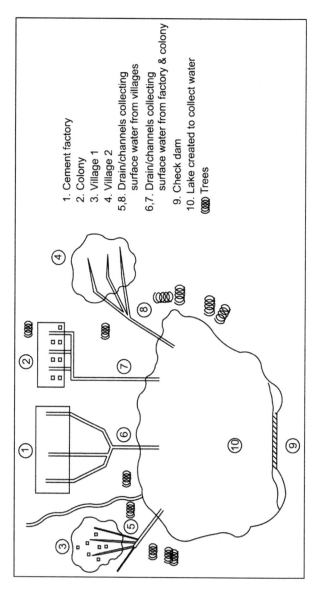

1. Cement factory
2. Colony
3. Village 1
4. Village 2
5,8. Drain/channels collecting surface water from villages
6,7. Drain/channels collecting surface water from factory & colony
9. Check dam
10. Lake created to collect water
⊗ Trees

Figure 6.3.1 Scheme of garland canals and check dam to collect rainwater for use of community and agriculture.

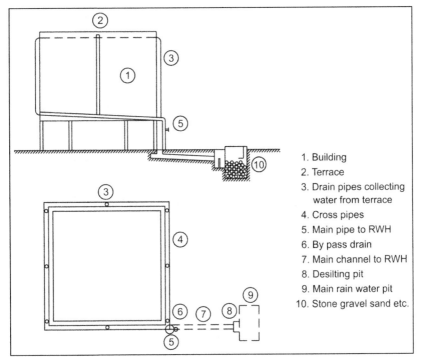

1. Building
2. Terrace
3. Drain pipes collecting water from terrace
4. Cross pipes
5. Main pipe to RWH
6. By pass drain
7. Main channel to RWH
8. Desilting pit
9. Main rain water pit
10. Stone gravel sand etc.

Figure 6.3.2 Rainwater harvesting scheme for a residential building.

3.8. Capital and maintenance costs of rainwater harvesting

Considering the total costs incurred by a cement plant during design and installation, the costs incurred by the RWH/water conservation systems mentioned above are insignificant.

ANNEXURE 1

Extracts from a letter from the Ministry of Environment & Forests stipulating conditions for environmental clearance for a cement plant project

The Ministry of Environment and Forests hereby accords environmental clearance to the above project under the provisions of EIA notification dated _____ subject to strict compliance to the following specific general conditions.

A. Specific conditions:

(i) The company shall comply with the conditions stipulated in the mining plan approval issued by the Indian Bureau of Mines and conditions of mining leases.

(ii) The gaseous and particulate matter emissions from various units shall confirm to the standards prescribed by the Pollution Control Board. At no time will particulate emissions from the cement plant including kiln, coal mill, cement mill, and cooler exceed 50 mgm/N m^3. Continuous on-line monitors for particulate emissions shall be installed. Interlocking facility shall be provided in the pollution control equipment so that in the event of the pollution control equipment not working, the respective unit(s) is shut down automatically.

(iii) Secondary fugitive emissions shall be controlled within the prescribed limits and regularly monitored. Guidelines/Code of Practice issued by the CPCB in this regard shall be followed. The company shall install an adequate dust collection and extraction system to control fugitive dust emissions at material transfer points. An atomized water spray system with reclaimer shall be installed in the silo used for the storage of ash. Storage of other raw materials shall be in closed–roof sheds. Covered conveyer belts shall be used to reduce fugitive emissions. Concreting of all the roads, and a water sprinkling system in the limestone and coal handling area shall be ensured to reduce fugitive emissions.

(iv) The proponent shall upload the status of compliance of the stipulated EC conditions, including monitored data, on their website and shall update the same periodically. It shall simultaneously be sent to the Regional Office of MoEF, the respective Zonal Office of CPCB and the SPC. The criteria pollutant, namely

SPM, RSPM, SO_2, NO_x (ambient levels as well as stack emissions) or critical sectoral parameters indicated for the project shall be monitored and displayed at a convenient location near the main gate of the company in the public domain.

(v) Ambient air quality including ambient noise levels shall not exceed the standard stipulated under the EPA or by the State authorities. Monitoring of ambient air quality and shall be carried out regularly in consultation with MPPCB and data for air emissions shall be submitted to the CPCB and SPCB regularly. The instruments used for ambient air quality monitoring shall be calibrated from time to time.

(vi) Efforts shall be made to reduce the impact of the transport of the raw materials and end products on the surrounding environment, including agricultural land.

(vii) The company shall make the efforts to utilize the high calorific hazardous waste in the cement kiln and necessary provisions shall be made accordingly. The company shall keep a record of the waste utilized and shall submit the details to the Ministry's Regional Office, CPCB, and SPCB.

(viii) Total groundwater requirement shall not exceed 1400 m^3/day for plant and mines. A copy of the permission letter shall be submitted to Ministry's Regional Office. The treated wastewater from STP and utilities shall be reutilized for green belt development and other plant-related activities, i.e., cooling and dust suppression in raw material handling area, etc., after necessary treatment. Zero discharge shall be strictly adopted and no effluent from the process shall be discharged outside the premises.

(ix) Rainwater harvesting measures shall be adopted for the augmentation of groundwater at the cement plant colony and mine site. In addition the company must also harvest the rainwater from the rooftops and storm water drains to recharge the groundwater. The company must also collect rainwater in the mined-out pits of captive lime stone mines and use the same water for the various activities of the project to conserve fresh water and reduce the river water requirement. The company shall construct the rainwater harvesting and groundwater recharge structures outside the plant premises also in consultation with the local Gram Panchayat and village heads to augment the groundwater level. An action plan shall be submitted to Ministry's Regional Office within three months from the date of issue of this letter.

(x)   The project proponent shall modify the mine plan of the project at the time of seeking approval for the next mining scheme from the Indian Bureau of Mines so as to reduce the area for external overburden dump by suitably increasing the height of the dumps with proper terracing. It shall be ensured that the overall slope of the dump does not exceed 28°.

(xi)   Topsoil, if any, shall be stacked with proper slope at earmarked sites only, with adequate measures, and shall be used for reclamation and rehabilitation of mined-out areas.

(xii) The project proponent shall ensure that no natural water course shall be obstructed due to any mining and plant operations. The company shall make the plan for protection of the natural water course passing through the plant and mine area premises and submit it to the Ministry's Regional Office.

(xiii) The inter-burden and other waste generated shall be stacked at the earmarked dump strictly and shall not be kept active for a long period. The total height of the dumps shall not exceed 30 m in three terraces of 10 m each and the overall slope of the dump shall be maintained at 28°. The inter-burden dumps shall be scientifically vegetated with suitable native species to prevent erosion and surface run off. Monitoring and management of rehabilitated areas shall continue until the vegetation becomes self-sustaining. Compliance status shall be submitted to the Ministry of Environment and Forests and its Regional Office on six monthly bases.

(xiv) The void left unfilled shall be converted into a water body. The higher benches of the excavated void mining pit shall be terraced and planting done to stabilize the slopes. The slope of higher benches shall be made gentler for easy accessibility to the water body by local people. Peripheral fencing shall be carried out along the excavated area.

(xv)   Catch drains and siltation ponds of appropriate size shall be constructed for the working pit and inter-burden and mineral dumps to arrest flow of silt and sediment. The water so collected shall be utilized for watering the mine area, roads, green belt development, etc. The drains shall be regularly desilted, particularly after monsoon, and maintained properly.

(xvi)   Garland drains of appropriate size, gradient, and length shall be constructed for both mine pit and inter-burden dumps and sump capacity shall be designed, keeping a 50% safety margin on above-peak sudden rainfall (based on 50-year data) and maximum discharge in the area adjoining the mine site. Sump capacity shall also provide an adequate retention period to allow proper settling of silt material. Sedimentation pits shall be constructed at the corners of the garland drains and desilted at regular intervals.

(xvii)   Dimensions of the retaining wall at the toe of inter-burden dumps and inter-burden benches within the mine to check runoff and siltation shall be based on the rainfall data.

(xviii)   Regular monitoring of groundwater level and quality shall be carried out by establishing a network of existing wells and constructing new piezometers at suitable locations by the project proponent in and around the project area in consultation with the regional director, Central Ground Water Board. The frequency of monitoring shall be four times a year: pre-monsoon (April/May), monsoon (August), post-monsoon (November), and winter (January). Data thus collected shall be sent at regular intervals to the Ministry of Environment and Forests and its Regional Office, the Central Ground Water Authority and the Central Ground Water Board.

(xix)   Blasting operations shall be carried out only during the daytime. Controlled blasting shall be practiced. The mitigating measures for control of ground vibrations and arresting fly rocks and boulders shall be implemented.

(xx)   The project proponent shall adopt wet drilling.

(xxi)   As proposed the green belt shall be developed in 33% in and around the plant as per the CPCB guidelines.

(xxii)   All the recommendations of the Corporate Responsibility or Environmental Protection (CREP) for the cement plant shall be strictly followed.

(xxiii)   Vehicular emissions shall be kept under control and regularly monitored. Measures shall be taken for maintenance of vehicles used in mining operations and in mineral transportation. The vehicles shall be covered with a tarpaulin and shall not be overloaded.

(xxiv)   Digital processing of the entire lease area using remote sensing technique shall be done regularly, once in three years, for monitoring the land use pattern, and the report submitted to the Ministry of Environment and Forests and its Regional Office.

(xxv)   A final Mine Closure Plan along with details of the Corpus Fund shall be submitted to the Ministry of Environment and Forests five years in advance of final mine closure, for approval.

(xxvi)   Necessary permission and recommendation of the State Forests department shall be obtained regarding impact of the proposed cement plant and mining on the surrounding reserve and protected forests and suggested conservation plans shall be implemented.

(xxvii)   The company shall comply with all the commitments made.

(xxviii)   Provision shall be made for the housing of construction labor within the site, with all necessary infrastructure and facilities such as fuel for cooking, mobile toilets, mobile STP, safe drinking water, medical health care, nursery, etc. The housing may be in the form of temporary structures to be removed after completion of the project.

B. General conditions

(i)   The project authority shall adhere to the stipulations made by the State Pollution Control Board (SPCB) and the State Government.

(ii)   No further expansion or modification of the plant shall be carried out without prior approval of this Ministry.

(iii)   At least four ambient air quality monitoring stations shall be established in the downwind direction as well as where maximum ground level concentrations of SPM, SO_2 and NO_x are anticipated in consultation with PCB/SPCB.

CHAPTER 4

Mining and Landscaping

4.1 Mining operations

For making 1 kg clinker, ~1.35 kg limestone is used. This is mostly obtained from captive limestone quarries.

Cement plants have mostly open cast mines. Limestone deposits come in many forms. Each deposit has its own characteristic in terms of physical and chemical properties. Each deposit has therefore to be obtained by employing the methods and machinery best suited to it.

4.1.1 Limestone deposits can occur in cultivated areas, in deserts, in forests and under the sea. Each location has its own general and specific problems which need to be taken into account when planning mining operations. The subject is discussed here with specific reference to maintaining or improving the ecology that existed before beginning the mining operations.

4.2 Overburden and contaminations

In open cast mining there is often overburden, consisting of topsoil, before limestone can be accessed. This overburden may be clay-like material that can be used along with excavated limestone as correcting material. In such a case there is no problem disposing of overburden. Similarly deposits may contain seams of contamination such as dolomitic materials, mica, etc., which need to be thrown out. In green mining, plans are made for disposal of unwanted material so that it does not grow into ugly heaps and spoil the landscape for years to come.

It is also necessary to uproot trees and shrubs to begin mining operations. In green mining, plans are made for greening a similar area at a suitable place or places by planting new trees that will over the years restore the ecology. It is sometimes possible to transplant fully grown trees and shrubs bodily at another place.

Topsoil can also be used to cultivate gardens at the cement plant and in the housing colony.

Designing Green Cement Plants
http://dx.doi.org/10.1016/B978-0-12-803420-0.00034-2
231

If planned from the beginning, handling overburden can be a manageable job without much extra effort.

4.3 Greening of slopes

As time goes on, mines go deeper and working faces become longer and are spread over considerable area. Excavated stone is brought to ground level for transport to the crushing plant. Maintenance of slopes is required to suit the equipment used, including shovels and dumpers. After a while the operation shifts to another face and what remains is denuded slopes of rock. Efforts can be made to green the slopes by covering them with topsoil and by growing grass and shrubs on them.

4.4 Groundwater

The depth of commercially mineable deposits is known from the mine's geological investigation report. Data on groundwater table in the area of the factory and mines should be obtained from appropriate sources. The two will give a fairly good idea of whether mining operations will expose underground water. Broadly speaking, mining is carried to a depth of four to five benches, each ~3 m in height. Groundwater table is not likely to be disturbed at this depth. However it can happen in some cases.

This water can then be used as the standby source of water and for the greening of mines and their surroundings.

4.5 Operation of quarries

Limestone has to be excavated from mines and taken to the plant for making cement. In that sense, this is not a sustainable operation. Over the years large excavated areas 15-25 m deep are created after stone has been removed. This is inevitable in any mining operation.

What can be done is to leave as small a "footprint" as possible by careful and planned mining operations. The general tendency is to exploit areas close to the plant or crushing plant to begin with and to progress further from the plant as time goes, and also to exploit high-grade deposits first.

While all other areas of operation make good use of computers and process automation to improve efficiency, mining operations lag somewhat behind.

Experts claim that major environmental problems occur due to bad mining practices. Therefore cement entrepreneurs desirous of putting up green cement plants have to pay greater attention to mining operations.

If the deposits are investigated in greater depth at the planning stage, mining plans can be made to simultaneously obtain high- and low-grade deposits and blend them.

Cement plants invariably install stacker reclaimer systems which have preblending facilities. Hence it stands to reason that the mining operations are carried out at two locations, one that contains high-grade stone and one that contains medium- or low-grade stone, which together can give the right mix for the type of cement being produced.

It means though that greater investment in mining equipment may have to be made to work on two locations simultaneously.

4.6 Mining in sensitive areas

It can happen that cement grade limestone deposits are discovered in "forest" areas and areas close to wildlife sanctuaries.

If there are growing markets for cement around these locations, these deposits attract entrepreneurs to exploit them for making cement. However the basic decision in this respect rests with the central and state governments of the country.

These days, therefore, the sanctioning authorities put in place a host of preconditions cement entrepreneurs must comply with regarding protection of the environment in sensitive areas before clearing the project.

These conditions need to be complied with. See Annexure 1 in Chapter 6.3.

4.7 Green belts

It is mandatory now to provide green belts around the plant and also between the plant and highway, railway track, etc. Besides cooling the environment, green belts of trees also serve as sound and dust barriers. Hence, creating green belts more than the mandatory number should be included in the planning and design of the plant and its surroundings.

Sanctioning authorities require that about 30% of the area acquired for the plant and housing colony should be left or made green.

If the site is in an area that has already many trees, this is not difficult. But in many cases cement companies have to make special efforts to plant trees, shrubs, and so on. Trees and shrubs take time to grow. Therefore this work needs to be taken up at the planning stage itself.

When the mines and the plant are located in a forest area, authorities insist on afforestation of an equivalent area. For that they must make available the area. If this is done, afforestation is not difficult. If authorities are not in a position to make equivalent area available, the project can be delayed.

4.8 Landscaping

Landscaping signifies creating esthetically pleasing and beautiful surroundings to make life more enjoyable. It is a conscious attempt to improve upon nature, making use of already existing features.

For example, the land acquired for the plant and colony may not be plain; it may have undulations and streams running through it. Every effort should be made to use existing topography to drain flood and rainwater. Advantage of the difference in ground levels should be used to reduce excavation and leveling efforts. For example, slopes may be conveniently used to orient the kiln and cooler. Differences in levels can be used to create esthetic and beautiful landscapes. A pleasant surrounding can boost the morale of the workforce in a big way.

Cement is gray; even green cement is gray. Raw materials used in cement have unattractive colors: coal is black. Raw materials are heaped in stockpiles all over the plant, adding to the drabness.

Greening of the surroundings softens the drabness and the plant blends with the surroundings rather than standing out as a sore thumb.

Gardens at the plant add beauty to it. Colonies look beautiful. With the latest pollution control equipment, it is possible to maintain ambient air quality and hence gardens and flowers remain bright and colorful whichever way the wind may blow.

4.8.1 There is another angle to this greening. That is conservation of water. Water to be used for this purpose should be treated gray water. Sanctioning authorities stipulate that effluent treatment plants are installed in all new projects to treat gray water.

They further suggest that the plants/trees to be grown should be drought resistant in type so that they do not make excessive demands on water.

4.8.2 Landscaping is a highly creative activity. No hard and fast rules can be laid down about it. While minimum norms can be established, there cannot be maximum norms.

4.9 Statutory stipulations regarding mine planning and operation and rainwater harvesting spelled out in Annexure 1 illustrate the governments's anxiety about protecting the environment and should therefore be looked upon as helpful guidelines for design and construction of green cement plants. Cement should keep them in mind at the design stage.

CHAPTER 5

Electrical Instruments and Appliances

5.1 Green electricity

Electricity generated using minimum natural and limited resources like fossil fuels can be called green electricity. We have discussed it in Section 5 on waste heat recovery.

Making efficient use of electricity also goes a long way to mitigating effects of global warming. Hence there is growing emphasis on the efficiency of electrical appliances of various kinds used in homes, industry, and business.

5.2 Bureau of energy efficiency

All over the world statutory and nonstatutory bodies now measure the efficiency of various electrical appliances and devices, including air conditioners and refrigerators, coolers, heaters, and so on, not leaving out the common light bulb or tube. It is recommended that only such appliances and instruments as have certificates of high efficiency issued by competent authorities are selected and installed in the cement plant.

5.3 CFL tubes

Introduction of CFL tubes for lighting has promised to reduce power consumption greatly and also emission of harmful gases to the atmosphere. However the flip side is the disposal of the tubes, which is problematic.

5.4 Efficiency norms

Efficiency norms help with selection of equipment and appliances that consume less energy and save power costs in times of increasing tariff rates. Further, indirectly there is overall less power consumption and hence less need to produce more.

Designing Green Cement Plants
http://dx.doi.org/10.1016/B978-0-12-803420-0.00035-4

Bureaus of energy efficiency give ratings to various appliances. They should be taken into account (using minimum of three stars) while selecting the appliances.

If transmission losses are reduced as well, existing power can go further and there will be fewer power shortage crises.

5.5 Use of air conditioners and refrigerators

CII Publications on green building design referred to in Annexure 1, Chapter 2 of this section gives guidelines regarding selection of refrigerating and air conditioning equipment. They should be followed.

5.6 Instrumentation, computerization and automation of operations

These help individually and collectively maintain operations continuously at optimum level. They improve availability of the plant and machinery. Various sections of the plant are interlinked. If they can be operated at their respective optimum levels, it would greatly enhance the operating efficiency of the plant as a whole and would also reduce GHG emissions and would thus promote sustainable development.

5.7 Thus by taking small steps in choosing equipment and electrical gadgets, significant contribution can be made to saving energy and thereby help keep the environment green.

CHAPTER 6

Renewable Sources of Energy

6.1 Renewable energy

This is the next step in designing green cement plants. The idea behind WHR is to generate power by using waste heat and thereby conserve fossil fuels. A logical next step is to examine alternative sources of energy: sources which are not exhaustible like fossil fuels and which do not contribute to GHG emissions. That will truly contribute to sustainable development.

Such sources are:

1. wind power
2. solar power
3. hydraulic power

These sources are obvious and yet largely underutilized, with the exception of hydraulic power.

6.2 Hydraulic power

Hydraulic power has been used on a small scale for centuries. It has also been used to generate power on a large scale. Geographic conditions are often the limiting factors. Hydraulic power stations cannot be installed anywhere and everywhere.

Hydraulic turbines are used to generate power. They need considerable areas of water. Hence hydraulic power systems are split. A reservoir is created by building a dam at a high level and water flows from it in pipes to turbines located at a much lower level. The potential energy of water is used to drive the turbine to which the alternator is coupled.

This is an inexpensive system of generation of power, requiring little maintenance. A reservoir receives/collects water every year in the rainy season. It also acts as a buffer to maintain the quantity of water required for generation year round. This has been a very reliable source of power for industry and cities.

Designing Green Cement Plants
http://dx.doi.org/10.1016/B978-0-12-803420-0.00036-6

6.2.1 In the mid fifties a large number of dams were constructed to bring large areas of land under cultivation in India. Hydraulic power stations were installed downstream to utilize the water from the dam to generate power before it was released into canals for irrigation. There was one problem though. Most rivers in North India had their origin in the Himalayas and brought silt with them. This affected power production over the years adversely.

A further development was low-head turbines which could generate power, albeit on a much smaller scale, from flow water in irrigation canals. Many small cement companies invested in such mini power plants and sold generated power to the grid in exchange for grid power to offset shortage in power supply.

The scope for installing hydraulic power plants by cement companies is however rather limited.

6.3 Wind power

Like hydraulic power, wind power is used on small scale to lift water from wells for agricultural purposes. Skylines in some European countries were dotted with windmills. (Don Quixote and his attack on windmills is well known.)

Presently wind farms consisting of a number wind turbines are designed to produce wind power on a much larger scale.

Many cement plants across the world have set up their own wind farms to produce electric power, which is then used by the cement plant itself or sold to the grid.

A cement plant may not be always located in an area suitable for setting up a wind farm on account of conditions like velocity of wind, its consistency day in and day out, and seasonal fluctuations. In such cases power generated by the wind farm will be sold out.

A wind farm occupies large areas. Such areas may not be available close to the cement plan.

The main problem with wind power is that winds change velocity and direction and this has a direct effect on the power generated. Further, some winds, like monsoons, are very much seasonal. They flow in one direction in some months and in the opposite direction in another. Wind power thus generated has to be stored so as to be available continuously.

6.4 Solar power

Solar power is being increasingly used for domestic purposes like heating water, cooking food, and so on. It is now used to generate electric power, with the development of photovoltaic cells (PV cells). Panels of PV cells are installed to catch the sunlight to the maximum possible extent during the day.

Like wind power, generation of solar power is not steady because of the movement of the sun. Intensity of light falling on the PV cells varies throughout the day. There is no generation of power in the night.

In daytime also, clouds can cut off the sun's rays for long periods. In areas with seasonal rains like monsoons generation of power is seriously affected in the rainy season for days at a time.

Panels have to be installed in the open. Winds deposit dust on them, thereby reducing their effectiveness. They also occupy lot of space.

6.5 Capacity factor and its implications

Both wind and solar power are variable due to variations in:
1. changes in speed and direction of wind on day to day as well as seasonal basis
2. changes in intensity of sunlight during the day, absence of sunlight at night; power actually generated and available varies considerably compared to installed capacity.

In other words, capacity factor is very low compared to that of thermal power.

Capacity factors of wind and solar power are given in Table 6.7.1 and 6.8.1 of Chapters 7 and 8, respectively, of this section.

Capacity factors vary from 20% to 40% for land-based wind power stations, 35% to 45% for offshore wind power stations, and 15% to 20% for solar power stations.

This means that to obtain a desired power output steadily on continuous basis it is necessary to install a much larger capacity. Capital costs therefore go up.

6.5.1 Typical capital costs for wind and solar (PV) plants presently are:

wind power land based   ~$1.6 million/MW
solar power (PV)   ~$4.0 million/MW

6.5.2 Let us say a cement plant requires 40 MW power.

It will arrange for grid power of 48 MW allowing a 20% factor of safety.

It will install a captive thermal power plant of 40% capacity, that is 16 MW.

It will also arrange to install a waste heat recovery plant of 20% capacity (8 MW).

It will install a wind power plant for 5% of its requirements, that is 2 MW.

For this, the actual plant to be installed would be 7 MW requiring a capital cost of $11.2 million.

If the capacity factor was 85% (as for a thermal power plant), capital costs required would be only $3.75 million.

6.5.3 Energy costs

For power plants based on renewable energy there are no fuel or operating costs (except maintenance costs). Their lifespan is also around 25 years. Therefore on a lifecycle basis (the levelized cost of energy) energy costs become attractive and viable. See Tables 6.7.1 and 6.8.1.

6.6 A careful and detailed exercise is required in each case to establish viability for different installed capacities to check capital costs and also to work out energy costs in Rs/$/kwh.

Since there are numerous sources of power, as listed above, overall energy costs have to be worked out for different permutations and combinations to arrive at optimum energy costs.

6.7 Technology is developing very fast for sources of renewable energy.

In the case of PV solar power, design of panels is changing, bringing down costs substantially. Therefore in the near future RE will compete with thermal power in capital costs and energy costs.

6.8 It should thus be possible to reduce consumption of fossil fuels by reduction in the demand for grid power. This will also reduce GHG emissions.

CHAPTER 7

Wind Power

7.1 General

Presently about 215,000 MW of wind power capacity exists the world over, with China leading the field. India, with a capacity of ~15,000 MW, is in the fifth position.

7.2 Wind farms

Large wind farms may contain several hundred turbines spread over hundreds of hectares. Wind power density is about 0.3 kw/sq.km.

The land on which wind turbines are erected can however be used for agricultural and other uses.

7.3 Wind speeds

Wind turbines require ~16 kmph or higher speeds, with steady non-turbulent winds, preferably the year round.

Wind atlases need to be checked to ascertain wind speeds and directions in different months of the year in any given region.

Wind speeds are higher at higher levels from the ground and also at higher altitudes. Wind turbines are generally installed at about 50 meters above ground level.

Ground surface also affects wind speeds. Therefore wind turbines over sea and water bodies are more efficient.

Doubling the altitude increases wind speed by 10% and power generated is proportional to V^3. Therefore a 10% increase in speed amounts to a 30% increase in power. See Fig. 6.7.1.

7.4 Planning for Installing Wind Turbines

1. Detailed data on wind conditions specific to the project site need to be obtained from **wind atlases** and from meteorological observatories. Conditions should be closely monitored for a year or more.

Designing Green Cement Plants
http://dx.doi.org/10.1016/B978-0-12-803420-0.00037-8

243

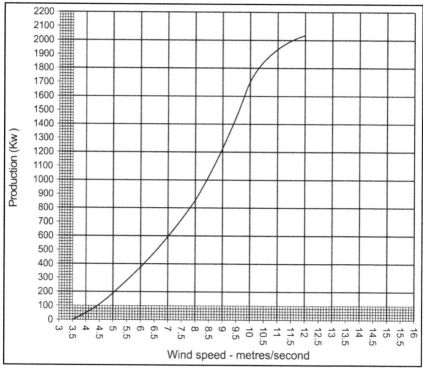

Figure 6.7.1 Relation between wind speed and power generated. *(Source: Brochure of Suzlon Energy Ltd).*

Detailed wind maps are constructed before decisions are made to set up wind farms.

2. Individual turbines are connected with a medium voltage of 35 kV. Land-based wind farms require installation of collector systems and substations to connect to the grid or to a captive user like a cement plant.

Offshore wind farms are less obtrusive as noise is mitigated by distance.

3. Wind farm developers may buy or lease the land; finance installation of turbines; and even operate and maintain them.

4. Coordination with local electricity boards is a vital factor if power generated is to be connected to the grid.

7.5 Cement plants themselves can set up wind farms and integrate them with the cement plant. The two need not be at the same place. The wind farm will be located at the site most suitable for generation of wind power and the cement plant will be located based on limestone deposit locations and market considerations. Power generated by the wind farm will be "wheeled" to the cement plant and/or the grid.

Power can flow in either direction. The farm and grid can supply power simultaneously to the cement plant. Or the farm can supply power to the grid. See Fig. 6.7.2.

Presently the capacity of a single wind turbine is about 850 kw. The number of turbines to be installed and the area required for the wind farm can be worked out for design capacity.

One turbine tower needs about 100 tons of steel.

7.6 Characteristics of wind power

As mentioned previously, the quantum of power generated by wind is not steady. It very much depends on the weather conditions. Actual outputs are unpredictable to some extent. Continuous use of only wind power requires complex optimizing procedures. Therefore, as a rule, wind power will be a part of the total power used in the plant.

Wind power saves fossil fuel, does not need water, and generates no waste.

7.6.1 Negative factors are large transport infrastructure and the manufacture, transport and installation of wind turbines.

7.6.2 There are complaints about the noise of wind turbines. If the wind farm happens to be in the migration routes of birds, there is a possibility of birds getting hit and killed.

7.7 Capital costs

Presently capital costs are on the order of 1.5 million US $ per MW of wind power for onshore installations and about 3.2–5 million $ per MW for offshore installations. Investment cost components vary for onshore and offshore wind farms. Payback period is short, about four to five months.

7.7.1 Operation & maintenance costs

Operation and maintenance (O & M) costs of wind power are about 1.2–2.3 US cents/kwh or 0.7 to 1.3 Rs. The levelized cost of energy at 50% capacity factor with a 10% discount is ~7 US cents/kwh or 3.9 Rs. Table 6.7.1 furnishes data on investment

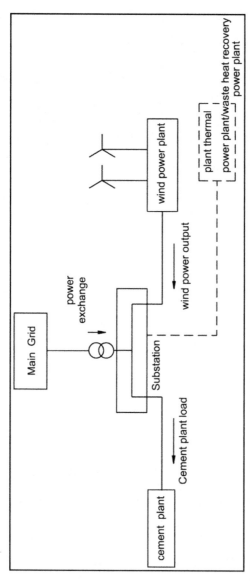

Figure 6.7.2 Simple scheme of grid-connected wind power plant supplying power to a cement plant. (*Source: Wind Generation Application for Cement Industry article by Mr. N. Miller and others*).

Table 6.7.1 Performance Data on Wind Power

Sr. No.	Type	Typical size (MW)	Investment cost ($/kW)	O&M cost USc/kwh	Capacity factor (%)	Design life time (years)	LCoE @ 10 % discount (USc/kwh)
1	On shore large	5–300	1200–2100	1.2–2.3	20–40	20	5.2–17
2	Off shore large	20–120	3200–5000	2–4	35–45	20	12–23

$ = US Dollar; USc = US cent; LCoE = levellised cost of energy.

Source: Annex III, Recent Renewable Energy Cost & Performance Parameters. Report by University of Cambridge.

costs, O & M costs, capacity factor, lifespan, and the levelized cost of energy (LCoE) for the two types of wind farms.

7.7.2 Wind power tariff

The wind power tariff is comparable to that of grid power at about 7 US cents or 4 Rs.

The grid will purchase power from the wind farm at 50% of its selling price in the market in peak hours and 40% in slack hours.

7.8 GHG emissions

Wind power is totally free of GHG emissions. A total of 1 MW of wind turbine power saves ~ 200 tons of CO_2 per year.

7.9 Operating installations of wind farms in cement plants

There are many installations of wind farms set up by cement companies either adjacent to the cement plant or elsewhere at suitable locations.

Plate 6.7.1 shows the typical wind farm of a cement plant.

Wind Park

Plate 6.7.1 Wind power for Tetouan Cement Plant. *(Source: CDM Simplified Design Project 2003).*

CHAPTER 8

Solar Power

8.1 General

Solar power is conversion of sunlight into electricity. There are two ways of doing this.

1. Concentrated Solar Power (CSP), in which sunlight is focused on an area containing water which is converted into steam and is used to generate power, as in a thermal power plant.

   CSP produces concentrated solar beam irradiation to heat liquid, solid or gas as in a regular TPS. The best sites for CSP are in equatorial belt cloud-free regions.

   Lenses or mirrors are used to focus rays onto a solar tower. Mirrors track the sun. CSP is more cost effective. These are observations in literature available on the system. One reason could be that – generation of electricity does not come to stand still when sunlight is not available because of its inherent thermal mass as it happens in case of PV cells. It is more efficient and has longer life.

2. PV cells, in which light is converted into electricity using photovoltaic cells (PV). Solar cells produce DC power, which fluctuates according to the intensity of irradiated light. This requires an inverter to produce power at the desired voltage frequency and phase. PV Systems are connected to the grid. They need batteries for backup.

   Fig. 6.8.1 shows a circuit diagram of a PV-cell-based solar power plant.

   PV cell electricity production globally in 2012 was ~55,000 MW; it is expected to increase by 2015 to 70,000 MW. India's share will be 5000 MW or more.

8.2 Solar power special features

Solar power is not available at night. The ability to store generated power is an important aspect of solar power. Available outputs must

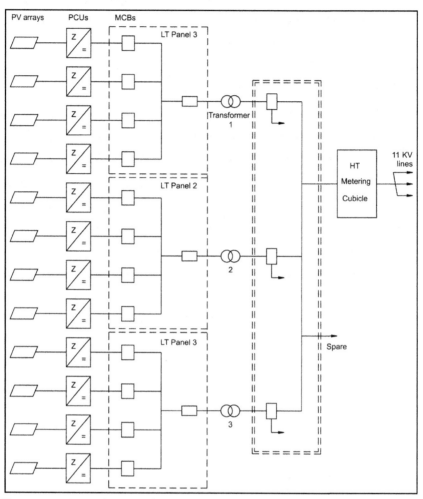

Figure 6.8.1 Circuit diagram of a PV solar power plant. *(Report IISc-DCCC 11 RE 1).*

be obtained when available and either stored for later use or transmitted to where it can be used.

Because of its inherent thermal mass, fluctuations in CSP systems are of less intensity.

8.2.1 Solar energy can be stored at high temperatures using molten salts.

8.2.2 Concentrated PV systems use concentrated light on PV cells. Many types of concentrators are available.

8.2.3 An inverter is required to convert it into AC at the desired voltage, frequency and phase.

8.2.4 Solar power does not have any solid, liquid, or gaseous byproducts. Its social & environmental impact is insignificant compared to a TPS.

8.3 Options for solar power

Basically there are several options for solar power installations.

1. Centralized
2. Distributed
3. On grid
4. Off grid

8.3.1 In grid-connected systems, excess power can be fed to the grid, with the grid serving as storage.

Inverters are needed to convert DC power to AC.

Off-grid systems have opportunities in areas of developing countries without electricity.

8.3.2 Distributed systems supply power to grid-connected consumers.

Centralized systems are common grids: they are not for any particular consumer.

8.4 Solar power in cement plants

Solar power can be used in any of the following ways:

1. solar PV cells
2. solar thermal
3. solar PV hybrid diesel.

Fig. 6.8.2 shows the three most common types of PV solar power systems.

8.4.1 Captive solar power systems in cement plants

There are three principal options:

1. standalone
2. grid tied
3. grid interactive with battery backup

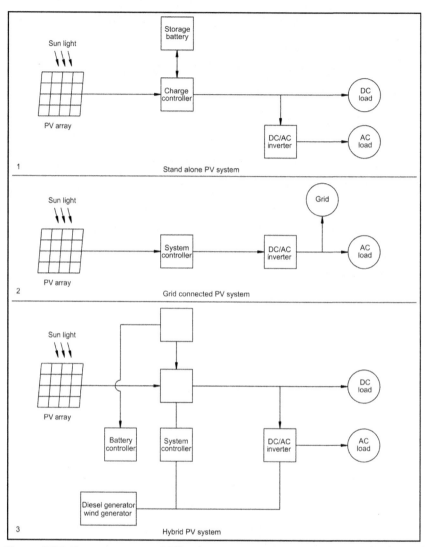

Figure 6.8.2 Common types of PV solar power systems. *(Science & technology of photovoltaics by P. Jayarama Reddy).*

Grid-tied systems feed power to the grid (whole or surplus); when power generation is off (sunlight not available) power can be bought from the grid.

This seems to be the most practical way of installing solar power. There is one more option, that of a hybrid system with wind power or a diesel generator (dg) set.

8.5 Layout of solar power station

Layouts differ radically depending on the system installed, that is,

1. concentrated solar power (CSP)
2. PV cells.

Plate 6.8.1 shows a CSP power plant.

8.5.1 A PV cell system requires considerable ground space, spread over hundreds of hectares.

Homes and offices may use solar power by installing PV cells on the rooftop or by integrating them in the architecture of the building.

8.5.2 In one installation of solar power for a cement plant, the footprint of the PV cells is 2650 sq. meters and is comprised of four arrays of PV cells, with each array containing 22 sets of 24 panels each.

8.6 Availability of solar power

Solar power, like wind power, is an intermittent source of power. All available power must be taken when available or sent where it can be used.

Neither sunshine nor its intensity is evenly spread on the surface of the earth.

Solar power thus cannot be used by itself on the large scale required by industry everywhere.

Plate 6.8.1 Concentrated solar power project. *(Solar power)*.

Tropical countries are more fortunate in that they have a maximum number of sunny days with high intensity of sunlight.

Metereological laboratories of various countries maintain data on solar radiation and daily duration of sunlight throughout the year.

In India, for example, the average number of sunny days is 250 to 300 in a year.

Annual global radiation varies between 1600 and 2000 kwh/sq. km. Equivalent energy potential is around 6000 gwh per year.

8.7 Capital costs

A 25-MW solar power plant is estimated to cost \$37.5 million, that is ≃\$1.5 million per MW.

Capital costs of solar power are high because substantial investment is required in inverters to convert DC power to AC.

Payback periods are between eight to twelve years currently. They are expected to come down considerably as costs of PV cells come down significantly due to advances in technology.

8.8 Costs of production

The cost of production of solar power is presently around 22 to 26 US cents per unit (Rs 12 to 14.5), compared to 7 to 9 US cents (Rs 4 to 5) per unit for grid power.

Solar power is cheaper on a lifecycle basis. The life of PV cells is 25 years or longer.

Maintenance and operational costs are much less.

Power from solar plants may be available in the future for 2.2 US cents (1.23 Rs), compared to grid power at 9 US cents (Rs 5).

8.8.1 Table 6.8.1 furnishes the current status of solar power systems by way of investment costs, operation and maintenance costs, capacity factor, life and LCoE.

8.9 Compared to wind power, solar power has a long way to go to become an alternative viable source of renewable energy in cement plants.

Table 6.8.1 Performance Data on Solar Power

Sr no.	Type	Typical Size (mw)	Investment cost ($/kw)	O & M Cost (USD/kwh)	Capacity Factor (%)	Design life time (years)	LCoE @ 10% discount (USc/kwh)
	PV Industrial fixed tilt	0.5-100	2700-5200	14-69	15-21	20-30	16-52

$ = US dollar; Usc = US cent; LCoE = levellised cost of energy.

Source: Annex III, Recent Renewable Energy Cost & Performance Parameters. Report by University of Cambridge.

RECOMMENDED READING

1. Green Buildings

1. Green Building Materials for Low Cost Housing by SP Agarwal, Head Organic Building Materials Division, Central Building Research Institute, Roorkee
2. IGBC Green Factory Buildings Rating System- Pilot Version 2009 Confederation of Indian Industries (CII) Sorabji Godrej Green Business Centre
3. IGBC Green Homes Rating System Version 1 by CII as above
4. Latest on Green Buildings by CII as above

2. Rain Water Harvesting

1. Texas Manual on Rain Water Harvesting Texas Water Development Board -third edition
2. RainXchange rainwater harvesting by Aqua space
3. Udai Presentation- Water Conservation Movements in India , Tarun Bharat Sangh

3. Renewable Energy - Wind & Soalr Power

1. Following references are from the same Publication
   - Chapter 1 Energy and Climate Change
   - Chapter 3 Direct Solar Energy
   - Chapter 7 Wind Energy
   - Chapter 9 Renewable Energy in the Context of Sustainable Development
   - Annex III Recent Renewable Energy - Cost & Performance Parameters
2. An analysis of opportunities in Wind Power Value Chain by Energy Alternatives in India (EAI)
3. Wind Farm
4. Wind Power in India
5. Energy for the future : Renewable Sources of Energy- White paper for a Community Strategy and Action plan 1997 European Comision
6. Renewable Energy
7. Wind Generation Applications for Cement Industry by - Nicholas Miller, Dilp Guru et al
8. Simplified Project Design Document for Small Scale Project Activities - Tetouan Cement Plant Clean Development Mechanism (CDM) 2003
9. Solar Power

Features of Large Green Cement Plants

Contents

List of Figures

List of Plate

List of Tables

List of Annexures

CHAPTER 1

Overview

1.1 The basic process of making green cement is the same as that used for OPC. For the various stages of manufacture the machinery used is also the same, subject to customizing the specific properties of materials and taking into consideration the size of the plant.

1.2 Limestone deposits

1.2.1 In the author's book **"Handbook for Designing Cement plants,"** in Chapter 5 of Section 3, ground rules for arriving at measured, indicated and inferred reserves and the area of mining for a 1-mtpa cement plant making OPC were explained.

Applying the same procedure, the mining lease required for a 10,000-tpd clinkering unit

making 3.5 mtpa of OPC would be:

Limestone deposits for	life of 30 years	220 million tons
Mining lease area		
(assuming 20-m average depth		550 ha
and in situ bulk density of 2 t/m^3)		
If provision is made for duplication		1100 ha

A 10% margin should be added for actual acquisition to allow for deterioration in quality of limestone at the boundaries.

1.2.2 If blended cements are made, the requirement of land for mining in relation to cement produced is reduced proportionately.

Mining lease area per million tons of cement would be:

For OPC	157 ha
For PPC	110 ha
For BFSC	63 ha

Designing Green Cement Plants
http://dx.doi.org/10.1016/B978-0-12-803420-0.00039-1

1.3 Capacities and sizes of main machinery and major auxiliaries

The process of arriving at the capacities of main machinery and corresponding auxiliaries in various sections of the plant remains the same as that outlined in Chapters 6-14 of Section 1, Basics, of the author's book **"Handbook for Designing Cement Plants."**

Since its publication, the common size of the cement plant has jumped to + 10,000-tpd clinkering capacity (single line) with a cement capacity of +3.5 mtpa for OPC and as high as +9.5 mtpa for slag cement (tpd = tons per day; mtpa = million tons per annum).

1.4 Machinery is sized to meet these capacities after applying the normal factors of design margin, conversion ratios, running hours per day, etc.

Tables are attached in Excel format to furnish data on:

Running hours	Table 7.1.1
Conversion/multiplying factors	Table 7.1.2

Using these data, sectional and hourly capacities have been calculated and are shown in **Table 7.2.1** in Chapter 2 for cement plants of 5000-, 7500- and 10,000-tpd cement kilns.

This table may be used to size various auxiliaries after applying appropriate factors. For example, a factor of ~20% would be added to size conveyors and feeders.

A machinery schedule for the total plant can be created section by section.

1.5 Capacities and sizes of machinery available

It is expected that with the growth in capacity of cement plants the machinery required to meet the required capacities in various sections will be developed. For example, a primary crusher working for six days a week and 12 h a day for a 10,000-tpd kiln would need to have a capacity of ~1900 tph.

Crushers of this capacity are in fact available.

The situation is more challenging if the plant capacity is to be duplicated at the same location.

Table 7.1.1 Running Hours in Various Sections of a Cement Plant

Sr. No.	Section	hr/shift	Shifts/day	hr/day	Day/week	hr/week	Day/year	hr/year	Remarks
	(A) Main plant								
1	Quarrying	6	2	12	6	72	312	3744	
2	Crushing	6	2	12	6	72	312	3744	
3	Stacker	6	2	12	6	72	312	3744	
4	Reclaimer	5	2	10	7	70	360	3744	Assuming supplies day's requirement of mill in 10 hr
5	Raw mill			20	7		360	7200	
6	Blending			20	7		360	7200	Works with raw mill
7	Extraction and kiln feed			24	7		330	7920	Works with kiln
8	Kiln			24			330	7920	
9	Coal stacker			6	7		360	2160	To suit capacity of wagon tippler and arrival of rake loads stacker reclaimer may have to work for thermal
10	Reclaimer			20	7		360	7200	Power station also may consider same
11	Coal mill			20	7		360	7200	Stacker reclaimer for main plant and 2nd unit
12	Cement mill			20	7		360	7200	
13	Packing and despatches	5	3	15	7		360	5400	

Continued

Table 7.1.1 Running Hours in Various Sections of a Cement Plant—cont'd

Sr. No.	Section	hr/shift	Shifts/day	hr/day	Day/week	hr/week	Day/year	hr/year	Remarks
						(B) Duplication			
1	Quarrying	6	3	17	6	102	312	5304	Same crusher in 3 shifts
2	Crushing	6	4	24	6	144	312	7488	2 crushers in 2 shifts
		6	3	17	6	102	312	5304	Same crusher in 3 shifts
3	Stacker	6	4	24	6	144	312	7488	2 crushers in 2 shifts
		6	3	17	6	102	312	5304	Same stacker
		6	4	24	6	144	312	7488	2nd stacker
4	Reclaimer	5	2	10	7	70	360	3600	Same reclaimer
		5	4	20			360	7200	Second reclaimer
5	Raw mill		4	40			360	14,400	2 raw mills
6	Blending			40			360	14,400	2 blending silos
7	Extraction, kiln feed			24				15,840	2 production lines
8	Kiln			24			660	15,840	2 kilns
9	Coal stacker			6	7	42	360	2160	To suit capacity of wagon tippler
10	Reclaimer			6	7	42	360	2160	
				10	7		360	7200	2nd stacker
				10	7		360	7200	2nd reclaimer
11	Coal mill			40			360	14,400	2 coal mills
12	Cement mill			40			360	14,400	2 cement mills
13	Packing and despatches	5	6	30			360	10,800	2 packing machines

Table 7.1.2 Conversion and Multiplying Factors

Sr. No.	Item	Unit	Design Margin	Conversion Ratio	Proportion	Overburden	Moisture	Loss in Transit	Margin on Design	Days Per Week	Total Multiplying Factor
1	Quarry raising	tpd	1.1	1.6	1	1.1	1.05	1	1	1.17	2.4
2	Limestone Wet, crushed,	tpd	1.1	1.6	0.9	1	1.03	1	1	1.17	1.91
3	Clay, wet	tpd	1.1	1.6	0.1	1	1.2	1	1	1	0.21
4	Sand,wet	tpd	1.1	1.6	0.06	1	1.1	1	1	1	0.12
5	Iron ore/laterite wet	tpd	1.1	1.6	0.02	1	1.15	1	1	1	0.040
6	Coal as fired	Cal. value sp. fuel con.		kcal/kg kcal/kg	650 700 750		**4000** 0.16 0.18 0.19	**4500** 0.14 **0.16** 0.17	**5000** 0.13 0.14 0.15	**6000** 0.11 0.12 0.13	
					Coal in kg/kg clinker						
				Select factor as applicable from above table							
7	Coal, wet coke breeze, wet	tpd tpd tpd	1.1 1.1 1.1 1.1 1.1 1.1	**0.14** 0.15 **0.16** 0.17 0.18 0.19	1 1 1 1 1 1	1 1 1 1 1 1	1.1 1.1 1.1 1.1 1.1 1.1	1.05 1.05 1.05 1.05 1.05 1.05	1 1 1 1 1 1	1 1 1 1 1 1	**0.18** 0.19 **0.20** 0.22 0.23 0.24

Continued

Table 7.1.2 Conversion and Multiplying Factors—cont'd

Sr. No.	Item	Unit	Design Margin	Conversion Ratio	Proportion	Overburden	Moisture	Loss in Transit	Margin on Design	Days Per Week	Total Multiplying Factor
8	Clinker	tpd	1.1	1	1	1	1	1	1	1	1.10
9	Gypsum, wet	tpd	1.1	0.053	1	1	1.15	1.05	1	1	0.07
10	Flyash	tpd	1.1	0.2	1	1	1	1.03	1.1	1	0.37
			1.1	0.3	1	1	1	1.03	1.1	1	0.00
			1.1	0.35	1	1	1	1.03	1.1	1	0.44
11	Blast furnace	tpd	1.1	0.4	1	1	1.1	1.03	1.1	1	0.55
	slag, wet		1.1	0.5	1	1	1.1	1.03	1.1	1	0.69
			1.1	0.6	1	1	1.1	1.03	1.1	1	0.82
			1.1	0.65	1	1	1.1	1.03	1.1	1	0.89
12	Cement opc	tpd	1.1	1.053	1	1	1	1	1.1	1	1.27
13	Raw meal	tpd	1.1	1.55	1	1	1.01	1	1.1	1	1.894
14	Kiln feed	tpd	1.1	1.55	1	1	1.01	1	1.1	1	1.894
15	Pulverized coal		1.1	**0.14**	1	1	1.02	1	1.1	1	**0.173**
		tpd	1.1	0.15	1	1	1.02	1	1.1	1	0.185

Sr. No.	Item	Unit	Design Margin	Conversion Ratio	Proportion	Overburden	Moisture	Loss in Transit	Margin on Design	Days Per Week	Total Multiplying Factor
16	Blended cements										
	(1) ppc										
	Flyash										
	20%		1.1	**0.16**	1	1	1.02	1	1.1	1	**0.197**
	30%		1.1	0.17	1	1	1.02	1	1.1	1	0.210
	35%		1.1	0.18	1	1	1.02	1	1.1	1	0.222
			1.1	0.19	1	1	1.02	1	1.1	1	0.234
	(2) Slag cement										
	Slag										
	30%		1.1	1.33	1	1	1	1	1.1	1	1.61
	40%		1.1	**1.54**	1	1	1	1	1.1	1	**1.86**
	50%		1.1	1.66	1	1	1	1	1.1	1	2.01
	60%		1.1	1.54	1	1	1	1	1.1	1	1.86
	65%		1.1	1.82	1	1	1	1	1.1	1	2.20
			1.1	2.22	1	1	1	1	1.1	1	2.69
			1.1	**2.86**	1	1	1	1	1.1	1	**3.46**
			1.1	3.33	1	1	1	1	1.1	1	4.03
17	Oil		1.1	0.08	1	1	1	1	1.1	1	0.10

Assuming that after duplication running hours for the crusher would be 15, crusher capacity to meet duplication would have to be 3000 tph.

For the major machinery, it is necessary to check the maximum capacities available from the respective producers.

It may be prudent to install multiple units so that loss of production due to breakdown is minimized.

1.6 Storage in stockpiles and silos

Cement plants need to arrange for stocks of raw materials and fuels as well as clinker and cements produced. Quantities stocked depend on the specific circumstances of each plant and have to be worked out in each case.

It is of course desirable to keep the investment in storages as small as practically possible.

1.6.1 Stockpiles

Crushed limestone and raw coal are stored in open or covered stockpiles. Limestone piles are generally linear. Coal piles can be either linear or circular. Piles occupy a lot of space. It is necessary to work out quantities to be stored and space needed in the layout not only for the first unit but also for the second production line of the same capacity.

It is also necessary to ascertain from various manufacturers the range of capacities of their stacker reclaimers.

Stacking capacities have to match crushing capacities and reclaiming capacities match raw material grinding capacities.

1.6.2 Blending silo with storage.

Universally continuous blending systems are now used to blend ground raw meal for firing in kilns.

Conventionally, 2-1/2 days' stock of raw meal is provided for. A 10,000-tpd kiln requires one 45,000-ton capacity or two 22,500-ton capacity blending with storage silos.

1.6.3 By applying this procedure, capacities for storage in various sections and dimensions of corresponding stockpiles and silos and a circular shed for clinker may be worked out.

1.7 Machinery schedule

A machinery schedule for the plant and its layout can now be planned and drawn. Though preliminary, it will serve as a good starting point. In a green plant various special points covered in the chapters on blending cements, alternative fuels and waste heat recovery are taken into account and provisions made to include machinery and space required for them in the layout.

1.8 Layout considerations

Principles of developing the plant layout and departmental layouts remain the same as previously described in the author's **"Handbook for Designing cement Plants,"** modified only to allow for machinery used and the scale of operations.

In case of green cement plants, layouts should provide for:
1. additional storage and handling capacities for blending materials and blended cements made;

2. storing and processing as required for feeding alternative fuels;

3. installing waste heat recovery system including power generation unit (in addition to thermal power plant).

Typical layouts developed as well as layouts of existing plants are included to illustrate the various points made above.

1.9 Process control, quality control and automation

There has been tremendous progress in the areas of process and quality control and automation, so much so that plants of 10,000-tpd capacity are manned by the same or even fewer people than a 3000-tpd plant used to be.

Quality control, beginning with collecting samples and ending with monitoring proportions of constituent materials and fuels in various stages of production, has been more or less fully automated.

So also are production processes. Use of simulators is now common to ensure continuous production at optimum level. Computerized control now covers all activities of cement production, beginning with quarrying and ending with dispatches and billing, and of course marketing, finance, and accounts.

CHAPTER 2

Sectional Capacities and Capacities of Major Individual Machines and Auxiliaries

2.1 Basis for arriving at sectional capacities

Three factors need to be considered in arriving at sectional capacities. They are the same whether the plant is large or small. They have been explained in detail in Chapters 8-10 of Section 1, Basics, of the **"Handbook for Designing Cement Plants."**

Tables 7.1.1 to 7.1.2 attached to Chapter 1 furnish data on running hours and conversion factors.

2.1.1 Overall multiplying factor

Using the data in the above tables, overall multiplying factors are determined and are shown in **Table 7.1.2**.

2.2 Sectional capacities for cement plants of 5000-, 7500- and 10,000-tpd capacity (kiln basis)

In this fashion the capacities for sections of the cement plant are worked out. Sectional capacities arrived at are shown in Table 7.2.1 for cement plants of 5000-, 7500- and 10,000-tpd capacity using a single kiln.

2.2.1 The trend is now to produce blended cements. The corresponding quantities for cements of different types in a year (OPC, PPC, and BFSC) will vary considerably. Table 7.2.2 shows the factors needed to calculate quantities of blended cements of different types produced from given clinkering capacities.

2.2.2 Attached nomogram Figs. 7.2.1-7.2.3 will help determine required capacities.

2.3 Capacities of major machines and auxiliaries in each section can now be worked out.

Designing Green Cement Plants
http://dx.doi.org/10.1016/B978-0-12-803420-0.00040-8

Table 7.2.1 Calculating Sectional Capacities for Cement Plants of Different Capacities

Sr. No.	Section	Item	Multiplying Factor	Hours Per Day	Factor for Hourly Capacity	Clinkering Capacity TPD		
						5000	7500	10,000
						Capacity in TPH		
1	Quarries	Stone raised	2.4	12	0.200	1000	1500	2000
2	Crushing	Limestone to be Crushed	1.91	12	0.159	796	1194	1592
		Crusher capacity	2.3	12	0.192	958	1438	1917
3	Stacker	Stacker (20% above crusher cap)	2.8	12	0.233	1167	1750	2333
4	Reclaimer	Reclaimer	1.89	10	0.189	950	1425	1900
	Raw material Grinding	Raw mill	1.72	20	0.086	430	645	860
5	Blending	Feed rate	1.89	20	0.095	475	715	950
		Extraction rate	1.89	24	0.080	400	600	800
6	Kiln	Feed rate	1.892	24	0.080	400	600	800
		Kiln, preheater Cooler	1.1	24	0.046	230	345	460
		Clinker conveyor	1.72	24	0.072	360	540	720

No.	Section	Item						
7	Coal grinding	Spillage	0.52	24	0.022	110	165	220
		Coal mill	0.2	20	0.010	50	75	100
		Coal firing	0.21	20	0.011	53	79	105
			0.22	20	0.011	55	83	110
			0.23	20	0.012	58	86	115
			0.2	24	0.008	42	63	83
			0.21	24	0.009	44	66	88
			0.22	24	0.009	46	69	92
			0.23	24	0.010	48	72	96
8	Cement Grinding	Cement mill						
		OPC	1.27	20	0.064	320	480	640
		PPC 30% ash	1.86		0.093	465	700	930
		BFSC 60% slag	3.46		0.173	865	1300	1730
9	Packing	OPC bagged	1.27	15	0.085	425	640	850
		Bulk	1.27	6	0.212	1060	1590	2120
		PPC bagged	1.82	15	0.121	605	910	1210
		Bulk	1.82	6	0.303	1515	2275	3030
		BFSC bagged	3.18	15	0.212	1060	1590	2120
		Bulk	3.18	6	0.530	2650	3975	5300

Table 7.2.2 Factors for Blended Cements

	Capacity in TPH				
A. PPC					
		Flyash (%)			
	20	25	30	35	
Clinker	75	70.0	65.0	60.0	
Flyash	19	23.75	28.5	33.2	
Gypsum	5	5	5	5	
PPC	100	100.0	100.0	100.0	
Ratio					
Cement/clinker	**1.33**	**1.43**	**1.54**	**1.67**	
B. Slag cement					
		Clinker (%)			
Clinker	65	55	45	35	30
Slag	30	40	50	60	65
Gypsum	5	5	5	5	5
	100	100	100	100	100
Ratio					
Cement/clinker	**1.54**	**1.82**	**2.22**	**2.86**	**3.33**
C. Composite Cement 1					
		Slag + flyash (%)			
Clinker	60.0	50.0	40.0	35.0	
Flyash	15	20	25	30	
Slag	20	25	30	35	
Gypsum	5	5	5	5	
	100	100	100	100	
Ratio					
Cement/clinker	**1.67**	**2.00**	**2.50**	**2.86**	
D. Composite Cement 2					
	Flyash + limestone (%)				
Clinker	70	65	60	55	
Limestone	5	5	5	10	
Flyash	20	25	30	30	
Gypsum	5	5	5	5	
Ratio					
Cement/clinker	**1.43**	**1.54**	**1.67**	**1.82**	

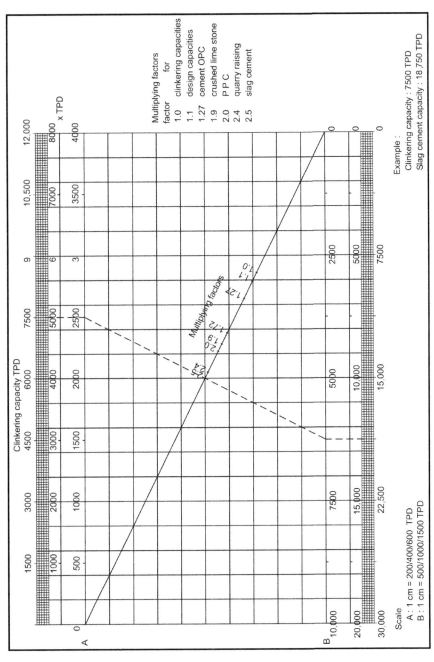

Figure 7.2.1 Pertinent capacities in TPD for various clinkering capacities.

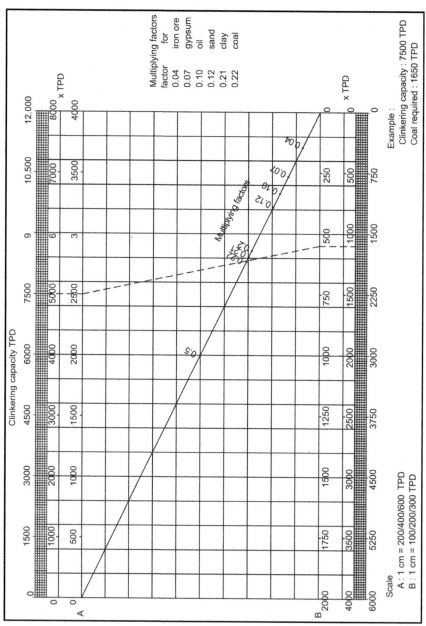

Figure 7.2.2 Pertinent capacities in tpd for gypsum, iron ore, clay, sand, etc.

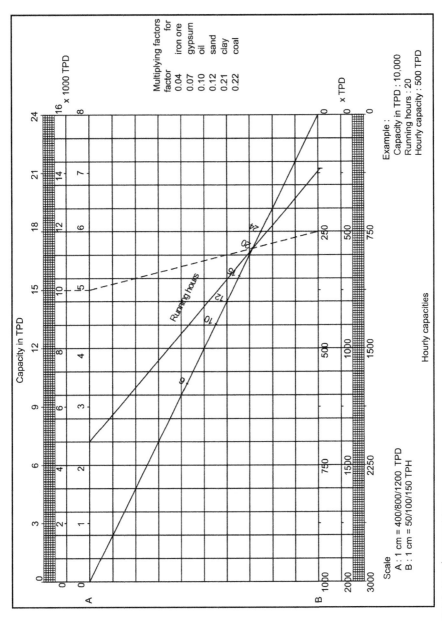

Figure 7.2.3 Hourly capacities.

Additional factors have to be added to basic sectional capacities for auxiliaries so that they can handle the flow of materials and gases under possible variations in production.

It is necessary to finalize system design by drawing a flow chart for each section.

2.4 A machinery schedule can then be created section by section for the whole plant (tpd = tons per day).

CHAPTER 3

Number of Units of Major Machines to Meet Sectional Capacities

3.1 Number of units of major machines to obtain required sectional capacity

Sectional capacities arrived at in Chapter 2 indicate the hourly capacity for the main machine to be installed in that section.

Ideally and most economically, this is a single machine large enough to yield the desired capacity.

3.2 Maximum sizes and capacities available

This has largely been possible as there has been corresponding development in the manufacture of cement-making machinery.

3.2.1 Crushing section

For a 10,000-tpd capacity, the sectional capacity required is 1600 tph and the capacity of the crusher is 2000 tph (see Table 7.2.1).

It must therefore be ascertained whether crushers of this capacity are actually available. If not, it will be necessary to install two crushers.

The second aspect to be considered is the possibility of doubling the plant capacity.

The obvious solution is to install a second crushing line of 1920-tph capacity. However if larger crushers are available, it may be economical to install a 2400-tph crusher to start with and to run it for fewer hours, and then run it for 17 h when plant capacity has been doubled.

This exercise is repeated for each section as each section has its special requirements.

Designing Green Cement Plants
http://dx.doi.org/10.1016/B978-0-12-803420-0.00041-X

3.2.2 Continuous blending and storage of raw meal

For example, when designing the continuous blending and storage section, the aspects to be considered are whether to have one or two silos to start with and whether this should be increased to two or three after duplication.

That will decide the capacity and dimensions of the blending silos.

3.2.3 Preheater, calciner, and kiln

While there is one kiln and one cooler, there are various possibilities with regard to the preheater and calciner to achieve capacities from 5000 to 10,000 tpd. Options are to install one or more preheater streams and one or more calciners. Up to a 5000-tpd capacity, currently one preheater and calciner are usually used.

Different machinery manufacturers have different design standards and their expertise should be considered.

3.2.4 Cement mills and cement storage silos

Cement plants are now making at least two types of cement, OPC and PPC or OPC and BFSC. Outputs from the same mills are different when making different types of cement. Mills have to be sized to make different types of cement and then used to make that type of cement (dedicated mills).

For the same clinkering capacity the quantity of cement produced is very different depending on whether it is OPC, PPC, or BFSC, or combinations in variable proportions.

The demand for different types in the area served by the cement plant decides the proportions of OPC and blended cements. The proximity of the market and seasonal fluctuations in demand decide the minimum quantities to be kept in stock, which in turn determines the numbers and capacities of cement silos.

3.3 The growth of the cement industry in terms of capacity and size of individual units has fortunately gone hand in hand with the development and growth of cement-making machinery. Availability of new designs has often given impetus to the growth of cement industry. Calciners and vertical mills are shining examples.

3.4 Whether it is possible to have one unit for different clinkering capacities is shown in Table 7.3.1. It is apparent that multiple units are required, mainly in cement grinding and dispatches sections.

Table 7.3.1 Number of Major Machinery Units in Each Section

Sr no	Section	Main Machine	Clinkering Capacity tpd — Capacity tph 5000	7500	1000	Clinkering Capacity tpd — Number of units 5000	750	10,000	Remarks
1	Quarrying					1	1	1	
2	Crushing	Primary crusher	1000	1500	2000	1	1	1	
			960	1440	1920	1	1	1	
3	Stacker	Stacker	1200	1750	2400	1	1	1	
	Reclaimer	Reclaimer	900	1500	1900	1	1	1	
4	Raw Mill		430	650	850	1	1	2	Optional
	Dust Collector					1	1	1	
5	Blending & storage	Continuous Blending silo				1	1	2	Optional
		Feeder	500	800	1000	1+1	1+1	2+1	1 stand by
6	Pyroprocessing	Kiln feed	400	600	800	1+1	2+1	2+1	1 stand by
		Preheater, no. of streams				1	2	3+1	Optional
		Calciner				1	2	3	Optional
		Kiln	5000 tpd	7500 tpd	10,000 tpd	1*	1*	2+1	Optional *Two support kiln **Three support kiln

Continued

Table 7.3.1 Number of Major Machinery Units in Each Section—cont'd

Sr no	Section	Main Machine	Clinkering Capacity tpd			Clinkering Capacity tpd			Remarks
			Capacity tph			Number of units			
			5000	7500	1000	5000	750	10,000	
7	Clinker conveyor to storage		350	525	700	1+1	1+1	2+1	1 stand by
	Clinker storage								
	Clinker conveyor from storage		400	600	800	1+1	1+1	2+1	1 stand by
8	Coal stacker					1	1	1	
	Coal reclaimer					1	1	1	
9	Coal mill		50	75	100	1	1	1	
10	Cement grinding	Cement mills							
		OPC	320	480	640	1	1	2	
		PPC 30% ash	460	700	920	1	2	2	
		BFSC 60% slag	800	1200	1600	2	3	3	Optional
11	Cement packing for dispatches by trucks & wagons	Packing machines							
		OPC 100% bagged	420	640	840	2	3	4	
		50% bagged	210	320	420	1	2	2	
		PPC 100% bagged	600	900	1200	3	5	6	
		50% bagged	300	450	600	2	3	3	
		BFSC100% bagged	1050	1600	2100	5	8	10	Optional
		50%bagged	525	800	1050	3	4	5	
					Capacities rounded off				

CHAPTER 4

Storages to be Provided

4.1 Materials to be stored and types of storages

A cement plant has to make arrangements for storing a host of bulk materials, like crushed limestone, additives, coal and gypsum. It also has to store semi-finished products like clinker, fly ash and slag.

It must store blended raw meal and the cements produced.

While limestone and coal are stored mostly in open stockpiles, clinker is stored either in covered conical sheds or in silos.

Additives, gypsum, and slag are stored in covered sheds.

Blended raw meal, cements, and fly ash are stored in silos.

The purpose of storage is to maintain production without interruption and to meet the variable market demands for cements at all times.

4.2 Investment in storage

Investment in storage is considerable in terms of capital costs. Further, stores occupy large space on the ground. Therefore it is desirable to keep storages to a minimum.

While plant operating experience indicates that it is useful to design storage after expansion, and this is generally done, this is not the case when designing a cement plant at a green field site. One can depend on norms followed by the cement industry in the past.

4.3 Conventional norms for storages

Table 7.4.1 shows conventions followed for storages of different materials and also shows quantities to be stored for cement plants of 5000-, 7500- and 10000-tpd capacity.

When duplicating, stocks can be reduced based on experience gained in the operation of the first unit.

For example, stock of crushed limestone may be reduced to 5 days instead of 7 days.

Designing Green Cement Plants
http://dx.doi.org/10.1016/B978-0-12-803420-0.00042-1

Table 7.4.1 Desirable Storage for Various Materials in Cement Plants of Different Capacities

Sr. no.	Material		Unit	1st Unit			After Duplication		
				Clinkering Capacity tpd			Clinkering Capacity TPD		
				5000	7500	10000	10000	15000	20000
1	Crushed limestone	factor		1.91	1.91	1.91	1.91	1.91	1.91
		daily reqmt	tons	9550	14325	19100	19100	28650	38200
		no. of day's stock		7	7	7	5	5	5
		storage	tons	**66850**	**100275**	**133700**	**95500**	**143250**	**191000**
2	clay	factor		0.21	0.21	0.21	0.21	0.21	0.21
		daily reqmt	tons	1050	1575	2100	2100	3150	4200
		no. of day's stock		30	30	30	20	20	20
		storage	tons	**31500**	**47250**	**63000**	**42000**	**63000**	**84000**
3	sand	factor		0.12	0.12	0.12	0.12	0.12	0.12
		daily reqmt	tons	600	900	1200	1200	1800	2400
		no. of day's stock		30	30	30	20	20	20
		storage	tons	**18000**	**27000**	**36000**	**24000**	**36000**	**48000**
4	iron ore/ laterite	factor		0.04	0.04	0.04	0.04	0.04	0.04
		daily reqmt	tons	200	300	400	400	600	800
		no. of day's stock		30	30	30	20	20	20
		storage	tons	**6000**	**9000**	**12000**	**8000**	**12000**	**16000**

Table 7.4.1 Desirable Storage for Various Materials in Cement Plants of Different Capacities—cont'd

Sr. no.	Material	Unit	1st Unit Clinkering Capacity tpd			After Duplication Clinkering Capacity TPD		
			5000	7500	10000	10000	15000	20000
5	coal wet raw							
	factor	tons	0.2	0.2	0.2	0.2	0.2	0.2
	daily reqmt		1000	1500	2000	2000	3000	4000
	no. of day's stock		20	20	20	14	14	14
	storage	tons	**20000**	**30000**	**40000**	**28000**	**42000**	**56000**
6	alternate fuels							
	factor	tons	0.04	0.04	0.04	0.04	0.04	0.04
	daily reqmt		200	300	400	400	600	800
	no. of day's stock		14	14	14	10	10	10
	storage	tons	2800	4200	5600	4000	6000	8000
7	clinker							
	factor	tons	1.1	1.1	1.1	1.1	1.1	1.1
	daily reqmt		5500	8250	11000	11000	16500	22000
	no. of day's stock		14	14	14	14	14	14
	storage	tons	**77000**	**115500**	**154000**	**154000**	**231000**	**308000**
8	gypsum							
	factor	tons	0.07	0.07	0.07	0.07	0.07	0.07
	daily reqmt		350	525	700	700	1050	1400
	no. of day's stock		30	30	30	20	20	20
	storage	tons	**10500**	**15750**	**21000**	**14000**	**21000**	**28000**

Continued

Table 7.4.1 Desirable Storage for Various Materials in Cement Plants of Different Capacities —cont'd

Sr. no.	Material		Unit	1st Unit			After Duplication		
				Clinkering Capacity tpd			Clinkering Capacity TPD		
				5000	7500	10000	10000	15000	20000
9	fly ash	factor		0.4	0.4	0.4	0.4	0.4	0.4
		daily reqmt	tons	2000	3000	4000	4000	6000	8000
		no. of day's stock		3	3	3	3	3	3
		storage	tons	**6000**	**9000**	**12000**	**12000**	**18000**	**24000**
10	slag	factor		0.82	0.82	0.82	0.82	0.82	0.82
		daily reqmt	tons	4100	6150	8200	8200	12300	16400
		no. of day's stock		2	2	2	2	2	2
		storage	tons	**8200**	**12300**	**16400**	**16400**	**24600**	**32800**
11	cement OPC	factor	1.27	1.27	1.27	1.27	1.27	1.27	1.27
		daily reqmt	tons	6350	9525	12700	12700	19050	25400
		no. of day's stock	7	7	7	7	5	5	5
		storage	tons	**44450**	**66675**	**88900**	**63500**	**95250**	**127000**
12	cement PPC	factor		1.82	1.82	1.82	1.82	1.82	1.82
		daily reqmt	tons	9100	13650	18200	18200	27300	36400
		no. of day's stock		3	3	3	3	3	3
		storage	tons	**27300**	**40950**	**54600**	**54600**	**81900**	**109200**

Table 7.4.1 Desirable Storage for Various Materials in Cement Plants of Different Capacities—cont'd

| | | | 1st Unit | | | After Duplication | | |
| | | | Clinkering Capacity tpd | | | Clinkering Capacity TPD | | |
Sr. no.	Material	Unit	5000	7500	10000	10000	15000	20000
13	cement BFSC							
	factor		3.18	3.18	3.18	3.18	3.18	3.18
	daily reqmt	tons	15900	23850	31800	31800	47700	63600
	no. of day's stock		3	3	3	3	3	3
	storage	tons	47700	71550	95400	95400	143100	190800
14	blended raw meal							
	factor		1.892	1.892	1.892	1.892	1.892	1.892
	daily reqmt	tons	9460	14190	18920	18920	28380	37840
	no. of day's stock		2.5	2.5	2.5	2.5	2.5	2.5
	storage	tons	23650	35475	47300	47300	70950	94600
15	pulverised coal *corresponding to 16 % coal consumption							
	factor		0.2★	0.2	0.2	0.2	0.2	0.2
	daily reqmt	tons	1000	1500	2000	2000	3000	4000
	no. of day's stock		0.3	0.3	0.3	0.3	0.3	0.3
	storage	tons	300	450	600	600	900	1200
16	oil							
	factor		0.1	0.1	0.1	0.1	0.1	0.1
	daily reqmt	tons	500	750	1000	1000	1500	2000
	no. of day's stock	8	8	8	8	6	8	6
	storage		4000	6000	8000	6000	9000	12000

Coal stocks may be reduced to 14 days instead of 20 days.

Stocks of cement may be reduced to 5 days instead of 7 days.

This will considerably reduce the investment to be made and space to be provided.

Considerable economies can be made in this way with regard to layout space and also capital costs.

CHAPTER 5

Developing Layouts for a Green Cement Plant

5.1. Developing a general plant layout for green cement plants

Principles of developing general plant layouts for green cement plants are the same as those for conventional plants. These have been dealt with in detail in Section 6 of the author's book "**Handbook for Designing Cement Plants**."

In the present context, the majority of new plants are large plants with clinkering capacities of 5000 tpd (single line) and annual capacities of +2.5 million tons. When designing layouts for such large plants, the possibility of future expansion also has to be kept in mind.

5.1.1. Size brings with it problems of logistics relative to handling and storing very large volumes and tonnages of raw materials like limestone, semi-finished product clinker, fuels, and finished products (cements).

5.2. The impact of introducing features that make a cement plant green

These have already been discussed. To summarize:

1. Provision for making blended cements to the maximum possible extent.

2. Using alternative fuels to the extent possible, taking into account availability on a continuous basis and the impact on design of machinery (such as providing kiln bypass and equipment for feeding tires), as well as provision for storage and treatment and feeding of alternative fuels.

3. Installation of waste heat recovery system and its integration with the captive thermal power plant.

4. In some cases installation of a power plant based on renewable energy like wind and/or solar power, if the proposed location is found suitable.

Designing Green Cement Plants
http://dx.doi.org/10.1016/B978-0-12-803420-0.00043-3

293

5. Facilities for rainwater harvesting/waste water treatment plant.

6. Designing green buildings, landscaping, green belts, etc.

5.3. Size, capacity, and space requirements for various items mentioned above have to be worked out and fit as well as possible in the general layout of the cement plant along with cement plant's main sections.

> **5.3.1.** The waste heat recovery plant and thermal captive power plant require a lot of auxiliaries, like a water treatment plant, cooling towers, and condensers, in addition to the main turbine and generator house; depending on the cycle to be used, Rankine or organic Rankine, a system for thermal fluids will also be needed to be included.

5.4. While layouts for blended cements and alternative fuels can be designed in house, the help of consultants and manufacturers of waste heat recovery systems should be requested to assess the location and space to be provided for installing these systems. The same is true with regard to wind and solar energy systems.

It is possible for wind and solar stations to be at different locations.

5.5. Space requirements for main equipment and storage

Main equipment

Once the capacities and types of main equipment are fixed, suppliers of this machinery will furnish departmental layouts of their respective machinery. If the machinery in different sections is bought from different suppliers, supplier layouts have to be fitted and dovetailed into an integrated layout. This can be done by the cement company or by consultants appointed by them.

Storage

As seen in the previous chapter on storage, huge quantities of different materials need to be stored. The next step is the ascertain the space provided in the layout for this storage in such a way that movement of materials between sections is efficient and economical.

Materials are stored in diverse ways: in silos, in open linear or circular stockpiles, in large-diameter conical shapes, and so on.

The attached Tables 7.5.1–7.5.7 furnish storage requirements in terms of space for cement plants of 5000- to 10,000-tpd clinkering capacity making OPC and blended cements.

> **5.5.1.** Fig. 7.5.1 is a nomogram for calculating the dimensions of stockpiles for limestone.

Table 7.5.1 Calculation of Volume of Linear Stockpile (Angle of Repose 35°) for Crushed Limestone

Part 1

Sr. No.	Width of Pile (m)	Height (m)	Area (m²)	Volume of Cone (m³)	St. Length in Meters							
					Volume of st. length in m³				Volume with cones in m³			
					150	200	250	300	150	200	250	300
1	20	7	70	725	10,500	14,000	17,500	21,000	11,225	14,725	18,225	21,725
2	30	10.5	158	2448	23,625	31,500	39,375	47,250	26,073	33,948	41,823	49,698
3	40	14	280	5803	42,000	56,000	70,000	84,000	47,803	61,803	75,803	89,803
4	45	15.75	354	8262	53,156	70,875	88,594	106,313	61,418	79,137	96,856	114,575
5	50	17.5	438	11,333	65,625	87,500	109,375	131,250	76,958	98,833	120,708	142,583

Part 2

Width	Total length of 1 pile				Total length of 2 piles in line			
	150	200	250	300	150	200	250	300
20	170	220	270	320	360	460	560	660
30	180	230	280	330	380	480	580	680
40	190	240	290	340	400	500	600	700
45	195	245	295	345	410	510	610	710
50	200	250	300	350	420	520	620	720

Continued

Table 7.5.1 Calculation of Volume of Linear Stockpile (Angle of Repose 35°) for Crushed Limestone—Cont'd

Part 3

		Calculation of stockpiles for crushed limestone for cement plants of different capacities			
		Plant capacity TPD			
		5000	7500	10,000	
Crushed limestone stock	tpd	66,850	100,275	133,700	Refer to Table 7.4.1
Bulk density	kg/m³	1600	1600	1600	
Vol.	m³	41,781	62,672	83,563	
Stock pile dimension Width × length		40 × 170	45 × 200	50 × 215	See part 1 above
Length of two piles in line	Meters	360	420	450	See part 2 above

Table 7.5.2 Calculation of Volumes of Straight Stockpiles of Trapezoidal Shape For Coal

Part 1

Sr. No.	Width at Base (m)	Height (m)	Area (m²)	Straight Length of Pile in Meters				Remarks
				100	200	300	400	
				Volume of stockpile in m³				
1	20	3	49.3	4930	9860	14,790	19,720	Angle of repose 40°
	30	3	79.3	7930	15,860	23,790	31,720	Height of piles 3 m
	40	3	109.3	10,930	21,860	32,790	43,720	
	45	3	124.3	12,430	24,860	37,290	49,720	
	50	3	139.3	13,930	27,860	41,790	55,720	

Volume of end portion neglected as being too small

Part 2

		Plant capacity tpd			
		5000	7500	10,000	
	Coal %	Stock coal tons			
	16	20,000	30,000	40,000	see Table 7.4.1
		Volume for bulk density 0.8 t/m³			
Vol.	m³	25,000	37,500	50,000	See part 1 above
Size of pile					
Width × length	m	45 × 200	45 × 300	45 × 400	
Length of two piles	m	420	620	820	

Table 7.5.3 Calculation of Volumes of Linear Stockpiles for Materials with Angle of Repose of 40°

Part 1

Sr. No.	Width of Pile (m)	Height (m)	Area (m²)	Volume of Cone (m³)	Volume of st. length in m³				Volume with cones in m³			
					St. Length in Meters				St. Length in Meters			
					150	200	250	300	150	200	250	300
1	20	8.4	84	880	12,600	16,800	21,000	25,200	13,480	17,680	21,880	26,080
2	30	12.6	189	2970	28,350	37,800	47,250	56,700	31,320	40,770	50,220	59,670
3	40	16.8	336	7040	50,400	67,200	84,000	100,800	57,440	74,240	91,040	107,840
4	45	18.9	425.3	10,024	63,775	85,050	106,313	127,575	73,811	95,074	116,336	137,599
5	50	21	525	13,750	78,750	105,000	131,250	157,500	92,500	118,750	145,000	171,250

Part 2

Width	Total length of 1 pile				Total length of 2 piles in line			
	St. length in meters				St. length in meters			
	150	200	250	300	150	200	250	300
20	170	220	270	320	360	460	560	660
30	180	230	280	330	380	480	580	680
40	190	240	290	340	400	500	600	700
45	195	245	295	345	410	510	610	710
50	200	250	300	350	420	520	620	720

Table 7.5.4 Sizing Continuous Blending and Storage Silos

Sr. No.	Item	Unit tpd	Kiln Capacity (tpd)		
			5000	7500	10,000
1	Raw meal/clinker	ratio	1.55	1.55	1.55
	Design factor		1.1	1.1	1.1
	Total factor		1.705	1.705	1.705
2	Raw meal per day	tons	8525	12787.5	17,050
3	storage capacity	No. of days	2.5	2.5	2.5
		tons	21312.5	31968.75	42,625
4	Bulk density aerated raw meal	t/m^3	0.96	0.96	0.96
5	Volumetric capacity of blending silo	m^3	22,201	33,301	44,401
6	Let h/d ratio be	Ratio	4	4	4
	Vol. of silo		$3.14 \times d^3$	$3.14 \times d^3$	$3.14 \times d^3$
	Diam. of silo	m	**19**	**21**	**23**
	Height	m	75	85	94
		Rounded to	76	84	92
	Allow for heap and escape of air	m	3	3	3
	Total height		**79**	**87**	**95**
	Height above floor	m	5	5	5
	Total height above ground	m	**84**	**92**	**100**
7	Volumetric capacity of blending silo	m^3			
		Rounded to	22,200	33,300	44,400
	h/d ratio		3	3	3
	Vol. of silo		$2.36 \times d^3$	$2.36 \times d^3$	$2.36 \times d^3$
	Diam. of silo	m	20	23	26
		Rounded to	20	23	26
	Height		**60**	**69**	**78**
	Allow for air space	m	3	3	3
	Height above ground		5	5	5
	Total height		**68**	**77**	**86**

Table 7.5.5 Calculation of Dimensions of Clinker Storage

		Part 1				
		Diam. of Storage in Meters				
Item	**Unit**	**60**	**70**	**80**	**90**	**100**
Area	m^2	2826	3847	5024	6359	7850
Angle of repose		35°				
Height of cone	m	21	24.5	28	31.5	35
Vol. of cone	m^3	19,782	31,413	46,891	66,764	91,583
	Height meter	**Vol. of cylinder**				
Vol. of cylinder in m^3	0	0	0	0	0	0
	1	2826	3847	5024	6359	7850
	3	8478	11,540	15,072	19,076	23,550
	5	14,130	19,233	25,120	31,793	39,250
	10	28,260	38,465	50,240	63,585	78,500
Total vol.	0	19,782	31,413	46,891	66,764	91,583
With cylindrical	1	22,608	35,260	51,915	73,123	99,433
Height in m^3	3	28,260	42,953	61,963	85,840	115,133
	5	33,912	50,646	72,011	98,557	130,833
	10	48,042	69,878	97,131	130,349	170,083

			Part 2		
			Clinkering Capacity tpd		
			5000	**7500**	**10,000**
Design factor			1.1	1.1	1.1
No. of days			14	14	14
Storage	tons		77,000	115,500	154,000
Bulk density	t/m^3		1.2	1.2	1.2
Vol.	m^3		64,167	96,250	128,333
		Rounded to	64,200	96,300	128,300
Alt. 1			Select size of clinker storage from part 1 above		
Let diam. $d =$	m		80	90	100
Cyl. height	m		3.45	5	5
		Rounded to	3.5		
Alt. 2					
Let diam. $=$	m		70		
Cyl. height	m		8.5		

Also see Fig. 7.5.2.

Table 7.5.6 Sizing Storage Silos for Cements

Sr. No.	Item	Unit	Factor	Clinker Capacity tpd		
				5000	7500	10,000
1	Cement produced 100% OPC	tpd	1.27	6350	9525	12,700
	Cement mill capacity	tph		318	476	635
	Cement stored	no days tons	7	44,450	66,675	88,900
	No. of silos			1	1	2
	Capacity each	tons		44,450	66,675	44,450
	Density cement	t/m^3		1.6	1.6	1.6
	Vol.	m^3		27,781	41,672	27,781
	h/d	ratio		3	3	3
	Diam. rounded to	m		**22**	**25**	**22**
	Height	m		**66**	**75**	**66**
	Height above ground	m		5	5	5
	Total height of top Rounded to	m		76	85	76
2	Cement produced 100% PPC	tpd	1.82	9100	13,650	18,200
	Cement stored	days	3	27,300	40,950	54,600
	No. of silos			1	1	2
	Capacity each	tons		27,300	40,950	27,300
	Bulk density	t/m^3		1.6	1.6	1.6
	Vol.	m^3		17,063	25,594	17,063
	Diam.	m		18.77	21.46	18.77
	Rounded to	m		19	21	19
	Total height of top	m		65.8	72.2	65.8
			Rounded to	66	72	66
3	Cement produced 100% BFSC	tpd	3.18	15,900	23,850	31,800
	Cement stored	days	3	47,700	71,550	95,400

Continued

Table 7.5.6 Sizing Storage Silos for Cements—cont'd

Sr. No.	Item	Unit	Factor	Clinker Capacity tpd		
				5000	7500	10,000
	Bulk density			1.6	1.6	1.6
	Vol.	m		29,813	44,719	59,625
	No. of silos			1	2	2
	Vol. each			29,813	22,359	29,813
	Diam.	m		22.57	20.52	22.57
	Rounded to			**23**	**21**	**23**
	Height	m		69	63	69
	Total height of top	m		**79**	**72**	**79**
			Rounded to			
4	Cement produced					
	50% OPC		1.27	3175	4763	6350
	50% PPC		1.82	4550	6825	9100
	Storage OPC	days	7	22,225	33,338	44,450
	Silos			1	1	1
	Storage for PPC	days	3	13,650	20,475	27,300
	Silos			1	1	1
	h/d ratio			3	3	3
	OPC silos					
	Vol.	$2.36 \times d^3$		13,891	20,836	27,781
	Diam.	m		17.54	20.05	22.05
	Rounded to	m		**18**	**20**	**22**
	Height	m		**54**	**60**	**66**
	Height to top	m		**63**	**69**	**75**
	Rounded to	m				
	PPC silos					
	Vol.	m^3		8531	12,797	17,063
	Diam.	m		14.93	17.07	18.77
	Rounded to	m		**15**	**17**	**19**
	Height	m		**45**	**51**	**57**
	Height to top	m		**53**	**59**	**66**
5	Cement produced					
	50% OPC	From	1.27	3175	4762.5	6350
	50% BFSC		3.18	7950	11,925	15,900
	Storage for OPC	tons	7 days	22,225	33337.5	44,450
	No. of silos			1	1	2
	Vol. each	m^3		13,891	20,836	13,891

Table 7.5.6 Sizing Storage Silos for Cements—cont'd

Sr. No.	Item	Unit	Factor	Clinker Capacity tpd		
				5000	7500	10,000
	storage for BFSC	tons	3 days	23,850	35,775	47,700
	Vol.	m³		14,906	22,359	29,813
	No. of silos			1	1	2
	Vol. each	m³		14,906	22,359	14,906
	Size of OPC silos			See 4 above		
	BFSC silo diam.	m		17.95	20.52	17.95
	Rounded to			18	21	18
	Height	m		54	63	54
	Total height to top	m		62.45	70.68	62.45
	Rounded to	m		62	71	62

Can also use nomogram no. 2-2-4 from author's book *"Nomograms for Design & Operation of Cement Plants."*

Table 7.5.7 Clinker Stored in Silos. Calculating Dimensions of Silos

Item	Unit	Clinkering Capacity tpd		
		5000	7500	10,000
Clinker stock 14 days	tons	77,000	115,500	154,000
Bulk density	t/m³	1.2	1.2	1.2
Clinker stock Vol.	m³	**64,167**	**96,250**	**128,333**
No. of silos		2	3	4
Clinker stock per silo	m³	32,083	32,083	32,083
h/d ratio		3	3	3
Diam. of silo	m	23.12	23.12	23.12
Rounded to	m	**23**	**23**	**23**
Height	m	**69**	**69**	**69**
Height to top	m	78.6	78.6	78.6
Rounded to	m	79	79	79
No. of silos		1	2	3
Vol. per silo		64,167	48,125	42,778
h/d ratio		3	3	3
Diam. of silo	m	29.06	26.43	25.42
Rounded to	m	**29**	**26**	**25**
Height		**87**	**78**	**75**
Height to top	m	97.8	88.2	85
Rounded to	m	98	88	85

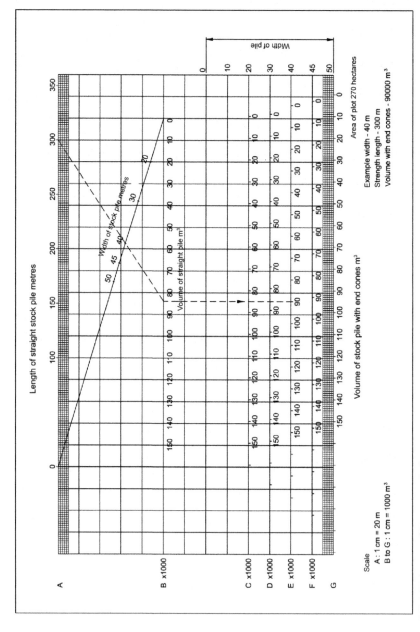

Figure 7.5.1 Volumes of linear stockpiles.

Fig. 7.5.2 is graph for calculating volumes of stockpiles for clinker.

5.6 Armed with this basic data development of the plant general layout can proceed.

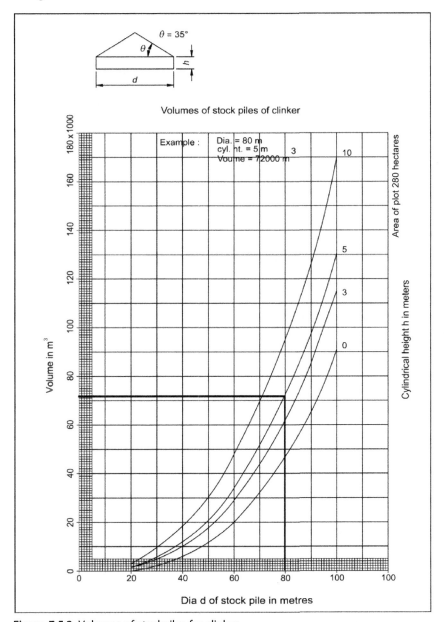

Figure 7.5.2 Volumes of stockpiles for clinker.

Two layouts for a clinkering capacity of 10,000 tpd to start with and with scope for duplication are attached for illustrative purposes.

The plant makes OPC and BFSC.

Basic data of the plant and the requirement calculations are furnished in Annexure 1.

See Figs. 7.5.3 and 7.5.4 for layouts, and Annexure 2, the legend for the layouts.

5.7 General plant layouts of existing cement plants

General plant layouts for existing cement plants made available by some cement companies and machinery manufacturers are attached with their permission. These endorse the various points made in this and earlier chapters.

See Plates 7.5.1–7.5.5.

ANNEXURE 1

Basic data and calculations for layouts 7.5.3 and 7.5.4
Developed for a 10,000-tpd clinker capacity
Cement plant making OPC and BFSC

Item	Unit	Quantity	Remarks	
Part 1: Cement mills, silos and packing and dispatches				
Capacity	tpd	10,000	clinker	1st unit
		10,000	clinker	2nd unit
Design margin	%	10		
			For margins and conversion factors see Table 7.1.2	
Cements made				
OPC	tpd	6930	Using 60% clinker	
Annually	tons	2,286,900	37	
BFSC	tpd	11,572	Using 40% clinker	
Annually	tons	3,818,760	62%	
		Rounded to	63	
Total		6,105,660		
	Rounded to	6,500,000	100%	
Cement mill capacities				
For OPC	tph	380		
Nos		1		
For BFSC	tph	650		
Number of mills		**2**		
Capacity each		325		
OPC to be stored	days	7		
	tons	53,200		
	Rounded to	50,000		
No. of OPC silos		1		
Size	Diam. × ht m	24 × 82		
BFSC to be stored	days	7		

Continued

Item	Unit	Quantity		Remarks	
	tons	91,000			
	Rounded to	100,000			
No. of BFSC silos		2			
Capacity each	tons	50,000			
Size	diam. × ht				
	m	24 × 82			
Cement dispatches		**OPC**		**BFSC**	
Per day total	tons	7000		13,000	
By road	65%	4550		8450	
By rail	35%	2450		4550	
Bagged	50%	3500		6500	
Bulk	50%	3500		6500	
Hourly bagging rate	tph	233		433	
	Rounded to	250		450	
Packing machines	no	**1**		**2**	
Bulk loading rate		**OPC**		**BFSC**	
Road	tph	152		282	
No. of loading points		1		1	
	Rounded to	150		300	
Bulk loading rate	tph	204		379	
Rail	Rounded to	200		400	
No. of loading points		1		2	
Rake loads Bulk and bagged cement		5		10	

Note: The main cement silo will be used for bulk loading by road. Cement will extracted and taken to packing machines.

One bulk loading platform for each type of cement will straddle a railway line. Cements will be transferred from main cement storage silos to a service silo as and when required. Its size will be 10 m diam. × 30 m high

Continued

Item	Unit	Quantity	Remarks	
Part 2: Coal and slag to be received				
1. Coal to be received for firing in kiln and calciner				
Capacity of kiln	tpd	10,000		
Wet coal	t/day	2000		
Days' stock		20		
Stock	tons	40,000		
Size of linear Stockpile	Width × length	**45 × 400**		
Two stockpiles	Meters	820		
2. Coal for captive power plant				
Avg. sp. power Consumption	kwh/t Cement	77		
Hourly production of cement	ton	925		
kwh		71,225		
mw		71.2		
After allowing for margins	mw	90		
Capacity of CPP at 40%	MW	36	captive power plant	
Assumed efficiency	%	40		
Coal required	tph	22 tph	cal. value 3500 kcal/kg	
Rounded to		25		
Wet coal	tph	29		
Per day	tons	700		
Stock to be maintained days	day	20		
	tons	14,000		
Size of stock pile	Width × length	**40 × 160**		
Two stockpiles in line	Meters	**340**		

Continued

Item	Unit	Quantity	Remarks	
Part 3: Stockpiles for crushed limestone				
Crushed stone	tpd	19,100		
Stock	days	7		
Stock	tons	133,700		
Size stockpile	width × length	**50 × 215**		
Length of 2 piles	meters	450		
Part 4: Continuous blending & storage silo				
Raw meal per day	tons	17,050		
Stock to be maintained	days	2.5		
	tons	42,625		
No. of silos		1		
Height/diam.		3		
Diam. of silo	m	25		
Height	m	85		
Part 5: Clinker storage				
Clinker	tpd	11,000		
Stock	days	14		
	tons	154,000		
Dimensions				
Clinker stock pile	Diam. × height meters	**100 diam. × 5 ht**		
Part 6: Sectional Capacities				
Quarrying	tph	2000	Nos.	
Crusher 1 stage	tph	1600	1	
Raw mill	tph	900	1	
Kiln	tpd	10,000	1	
Preheater six-stage		2 stream	1	
Calciner		offline	1	
Clinker cooler		10,000	1	
Coal mill	tph	100	1	

Continued

Item	Unit	Quantity	Remarks
Part 7: Alternative fuels			
Calorific value	kcal/kg	4000	
Sp. fuel con.	kcal/kg	750	
quantity	%	20	
To be fired in		calciner	
rate	tph	20	
af per day	tons	480	
	say	500	
Stock days		14	
	tons	7000	
Part 8: Slag for making BFSC			
Clinker used to make BFSC	tpd	4000	
Slag required	tpd	6000	
With margin	tpd	6600	
Wet slag	tpd	8500	
Storage wet slag	days	2	
	tons	17,000	
Storage dry slag	days	2	
	tons	15,500	
Part 9: Captive power plant & waste heat recovery plant			
Plant	Main machinery, auxiliaries, and electrical switch gear, condensers cooling towers, etc., as required		
Waste heat recovery system	Capacity 15 MW operating on Organic Rankine cycle		
	Complete with main machinery and auxiliaries as required for CPP and systems for thermal and working fluids		

Note: Refer to Tables 7.2.1, 7.3.1, and 7.5.1–7.5.7 for the basis of quantities and sizes cited above.

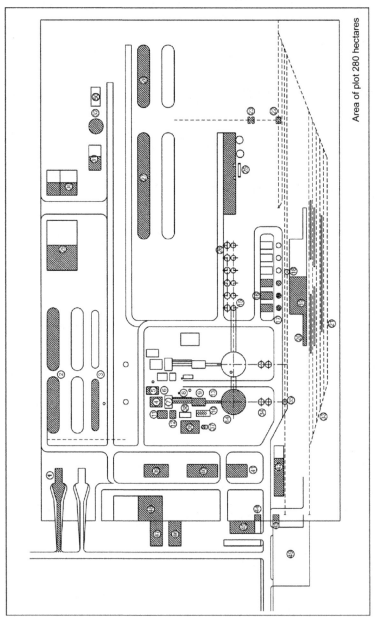

Area of plot 280 hectares

Figure 7.5.3 General plant layout for a 10,000-tpd cement kiln for OPC and BFSC.

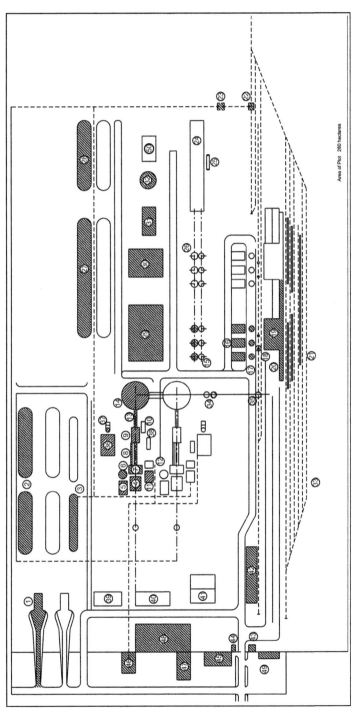

Area of Plot 280 hectares

Figure 7.5.4 General plant layout for a 10,000-tpd cement kiln for OPC and BFSC.

ANNEXURE 2

Legend for General Plant Layouts (Figs. 7.5.3 and 7.5.4)

For a 10,000-tpd clinkering unit making OPC & BFSC

Note: Basis and calculations are as per Annexure 1.

The first unit is shown shaded. Duplication is shown without shading.

Item No.	Legend	Remarks
1	Crusher	
2	Stacker Reclaimer System for crushed limestone	Two linear piles
3	Storage for additives and correcting materials	Covered shed
4	V R Mill for grinding raw materials	
5	Dust Collector for raw mill and kiln	
6	Continuous Blending and Storage Silo	
7	Two-stream six-stage preheater & calciner	
8	Rotary kiln	
9	Clinker cooler	
10	ESP for cooler	
11	Coal mill	
12	AF firing for calciner	
13	Clinker conveyor	
14	Clinker storage	
15	Hoppers for Cement Mill	
16	Cement Mills VRMs	1 for OPC, 2 for BFSC
17	Cement Silos	1 for OPC, 2 for BFSC
18	Silo for bulk loading by rail	1 each for OPC & BFSC
19	Packing machines	1 for OPC, 2 for BFSC
20	Truck-loading facility for bagged cement	
21	Wagon-loading facility for bagged cement	3 platforms for 1st unit
22	Wagon tippler for coal and slag	
23	Coal crusher	
24	Storage for wet and dry slag	
25	Slag dryer	
26	Slag hoppers	
27	Stacker Reclaimer for coal for cement plant	Linear piles in line

Item No.	Legend	Remarks
28	Stacker Reclaimer for coal for captive power plant	Linear piles in line
29	Captive Power Plant	
30	Condensers	
31	Cooling towers	
32	Overhead water tank	
33	BG railway siding	
34	Silos for extra clinker	
35	Silo for wagon loading of clinker	
36	WHR System	Common for preheater and cooler
37	Condenser cooling tower for WHR	
38	Central Control Room	
39	Engineering Offices	
40	Laboratory and Quality Control	
41	General Stores	
42	Open Store Yard	
43	Security	
44	Time Office	
45	Main Administrative Office	
46	AF Store and Processing	
47	Grid Substation	
48	Main Substation of the plant	
49	Yard for trucks/bulk carriers	
50	Cooling pond	

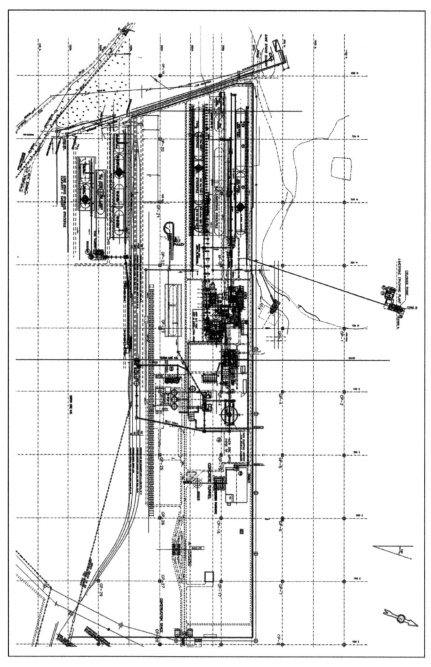

Plate 7.5.1 General layout of a cement plant with a 13,000 tpd kiln.

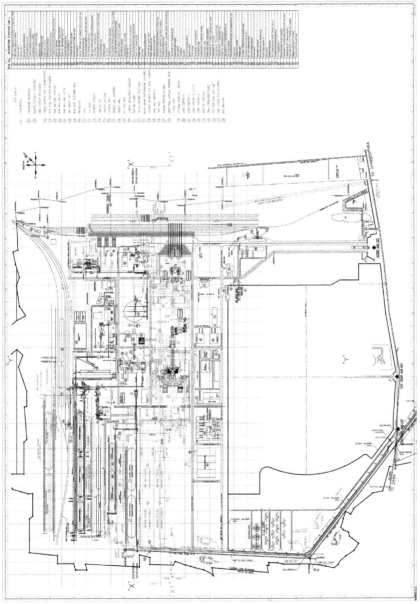

Plate 7.5.2 Layout of a cement plant with 8000-tpd Cement Plant.

S.NO.	DESCRIPTION
1.	PREBLENDING STOCKPILE— CAP.:-2x 60,000 T
2.	SILICA STOCKPILE— CAP.:-5,000 T
3.	COAL STOCKPILE— CAP.:-6x 4,500 T
4.	RAW MILL HOPPERS
5.	RAW MILL
6.	RAW MILL BAG HOUSE / SUB STATION
7.	BLENDING SILO— CAP.:-30,000 T
8.	PREHEATER
9.	KILN
10.	T.A DUCT
11.	COOLER E.S.P.
12.	CLINKER SILO— CAP.:-1,20000 T
13.	REJECT SILO
14.	DUMP HOPPER FOR LATERITE AND SILICA
15.	DUMP HOPPER FOR LIGNITE
16.	CAPTIVE POWER PLANT
17.	WORK SHOP (MECH. & ELECT.)
18.	GENERAL STORE
19.	STORE YARD
20.	LIGNITE MILL HOPPER
21.	LIGNITE MILL
22.	C.C.R/Q.C. /Q.C/Q.C.P BUILDING
23.	KILN SUB. STATION
24.	PERSONNAL & TIME OFFICE
25.	ADMINISTRATION OFFICE
26.	SECURITY OFFICE
27.	SHED FOR FIRE FIGHTING DEPT.
28.	CAR / SCOOTER & CYCLE PARKING
29.	WORKERS CANTEEN
30A.	OFFICERS CANTEEN
31.	MAIN GATE
32.	MATERIAL GATE FOR TRUCK MOVEMENT
33.	WEIGH BRIDGES
34.	CEMENT DISPATCH OFFICE
35.	PETROL & DIESEL PUMP
36.	RAW WATER TANK FOR CEMENT PLANT
37.	BLACK START DG SET
38.	DESALTINATED WATER TANK
39.	SOFT WATER TANK
40.	SEDIMENTATION TANK
41.	RECIRCULATION WATER TANK
42.	PRIMARY SCREEN & CRUSHER FOR LIMESTONE
43.	SECONDARY SCREEN FOR LIMESTONE
44.	BULK LOADING
45.	COMPRESSOR HOUSE FOR PYRO
46.	PUMP HOUSE FOR PYRO SECTION
47.	COMPRESSOR ROOM
48.	RAILWAY TRACK (TOP OF RAIL)

LEGEND

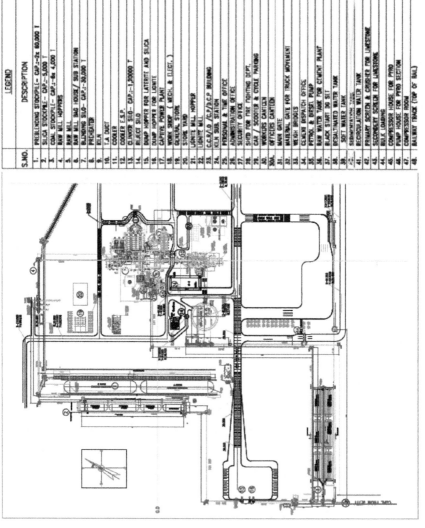

Plate 7.5.3 Layout of a 6.7-mtpa-capacity modern cement plant (clinkering unit).

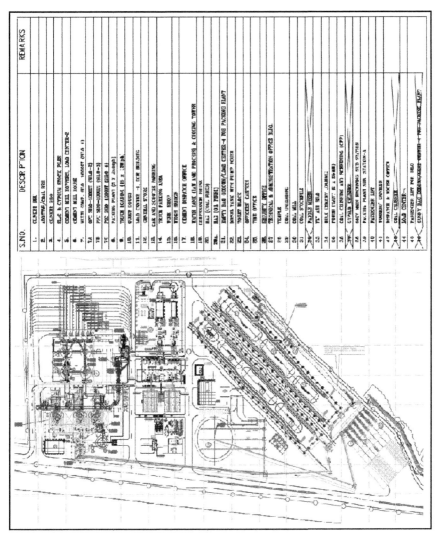

S.NO.	DESCRIPTION	REMARKS
1.	CLINKER SILO	
2.	ADDITIVE/COAL ROD	
3.	CLINKER SILO	
4.	SLAG & GYPSUM STACK PILES	
5.	CEMENT MILL HOPPERS, LOAD CENTER-2	
6.	CEMENT MILL HOUSE	
7.	WITH COMP. AREA HOPPERS (WFA 1)	
7A.	OPC SILO-1000MT (SILO-2)	
7B.	PPC SILO-1000MT (SILO-3)	
7C.	PSC SILO 1000MT (SILO 4)	
8.	PACKING PLANT (N 3 nos/ph)	
9.	TRUCK LOADING (10 X 10MVA)	
10.	QUEST DOORS	
11.	LOAD CENTRE -1, DCR BUILDING	
12.	GENERAL STORE	
13.	CAR AND SCOOTER PARKING	
14.	TRUCK PARKING AREA	
15.	RAIL SHED?	
16.	WEIGH BRIDGE	
17.	CEMENT DISPATCH OFFICE	
18.	WATER TANK (LAT AND PROCESS) & COOLING TOWER	
19.	COMPRESSOR HOUSE	
20.	BAG (COAL SHED)	
20A.	BAG (OIL TUBE)	
21.	EMPTY BAG GODOWN/LOAD CENTER-4 FOR PACKING PLANT	
22.	DIESEL TANK WITH PUMP HOUSE	
23.	TOILET BLOCK	
24.	OFFICERS CANTEEN	
25.	TIME OFFICE	
26.	SECURITY OFFICE	
27.	TECHNICAL & ADMINISTRATION OFFICE BLDG.	
28.	TEMPLE	
29.	COAL UNLOADING	
30.	COAL MILL	
31.	COAL STOCKPILE	
32.	PADDLE FEEDER	
33.	FLY ASH SILO	
34.	BULK LOADING LOADING	
36.	POWER PLANT (6 x 25+MW)	
36.	COAL CRUSHING AND SCREENING (CPP)	
38.	GYPSUM CRUSHER	
38.	66KV WIDE RECEIVING STD STATION	
39.	PACKING PLANT SUB STATION-A	
40.	PASSENGER LIFT	
41.	RUNNING CRUSHER	
42.	ELEVATOR & REVERSE CONVEY	
43.	COAL CRUSHER	
44.	LOAD CENTER-?	
45.	PASSENGER LIFT FOR SILO	
	CONY BAG DEGUMMAGED HOPPER-1 FOR PACKING PLANT	

Plate 7.5.4 Layout of cement grinding, packing, and dispatching section of plant in Plate 7.5.3.

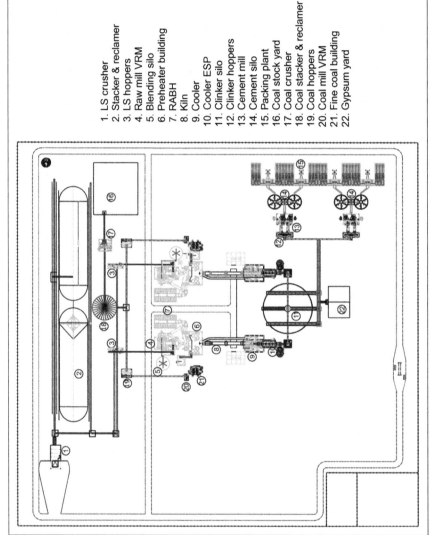

1. LS crusher
2. Stacker & reclamer
3. LS hoppers
4. Raw mill VRM
5. Blending silo
6. Preheater building
7. RABH
8. Kiln
9. Cooler
10. Cooler ESP
11. Clinker silo
12. Clinker hoppers
13. Cement mill
14. Cement silo
15. Packing plant
16. Coal stock yard
17. Coal crusher
18. Coal stacker & reclamer
19. Coal hoppers
20. Coal mill VRM
21. Fine coal building
22. Gypsum yard

Plate 7.5.5 General layout of a 2.0-mtpa cement plant.

CHAPTER 6

Layout of Dispatches of Cement Bagged and in Bulk

6.1 General

Sending cement(s) produced to the market(s) in the forms and modes demanded poses several challenges in the design of packing plant and bulk dispatch and truck- and wagon-loading facilities.

6.1.1 It is also necessary to shape the layout to fit a railway siding which has its orientation fixed at its origin.

Besides transporting bagged/bulk cement by rail, it also brings in slag, coal, and other materials.

6.1.2 External factors decide the shape and size of the plot of land available for factory and siding.

A lot of ingenuity is required to make the best of facilities available.

6.2 Variable factors

The principal variable factors are:

(1) types and quantities of cements made
(2) whether production and dispatches of various types are simultaneous
(3) modes of dispatches
    (1) bulk and bagged and proportions thereof for different cements
    (2) proportions sent by road and by rail of each type and of bulk/bagged
    (3) palleting and container mode for export

6.3 Facilities to be installed

(1) bagging machines with auxiliaries like automatic bag feeding machines

(2) truck- and wagon-loading machines for bagged cement

(3) bulk loading facilities with weighing arrangements for transport by road and rail simultaneously.

6.4 Thus the layout beginning with storage of cement and ending with dispatches of cement is divided into three broad sections.

(1) simultaneous storage and withdrawal of cement(s)

(2) bagging operation/bulk loading

(3) dispatches in bulk or bagged, by road and/or rail.

6.5 The market dictates the types of cement and the demand for them now and in the near future. The location of the cement plant in relation to the market determines the time to fulfill a cement order. In turn this decides what quantities of cement are stored. These storages are the buffer between production and dispatches.

6.6 Silos are designed so that bulk loading operations are carried out right under the silo so that truck transport is simultaneous with withdrawal for bagging of cement.

6.6.1 In smaller plants the silo may be sectioned to store more than one type of cement, which is then withdrawn simultaneously for further processing. See Plates 7.6.1 to 7.6.3.

6.6.2 Some silos are earmarked for bulk loading by rail in special wagons. This is particularly useful for plants exporting cement.

6.6.3 Sometimes cement is exported in bags. This may call for palleting of bags,

loading pallets into containers and transporting loaded containers to ships by rail and/or road.

Because of so many variables, there is no standard layout for this last phase of activity in a cement plant.

The layout is worked out to meet the specific needs of a given cement plant.

6.7 Principles of design remain the same as those outlined in the **"Handbook for Designing Cement Plants."** As the scale of operations is much larger, the area is correspondingly larger.

6.8 This causes problems with effective supervision. Various operations are now automated, for example:

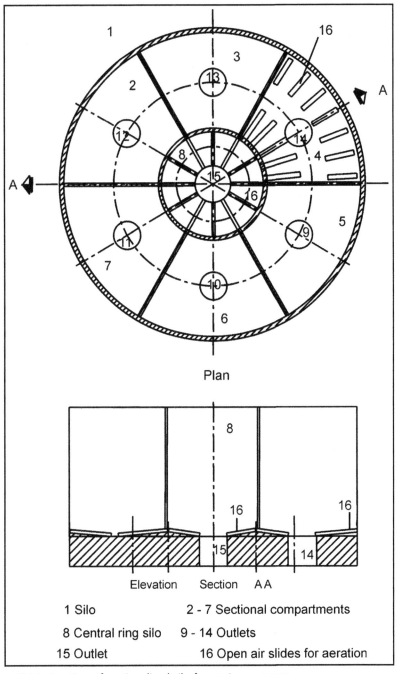

Plan

Elevation   Section   A A

1 Silo   2 - 7 Sectional compartments
8 Central ring silo   9 - 14 Outlets
15 Outlet   16 Open air slides for aeration

Plate 7.6.1 Aeration of sectionalized silo for storing cements.

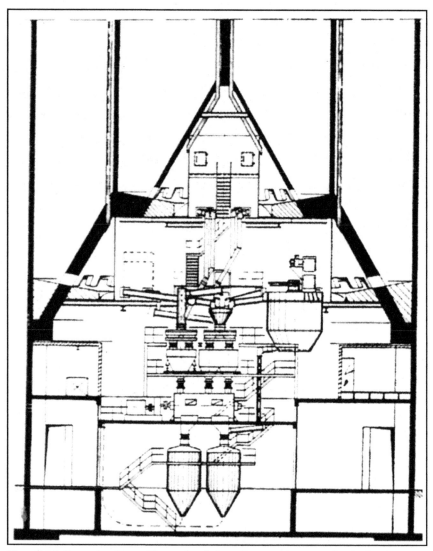

Plate 7.6.2 Extraction from different levels in multi-compartment silos for loading and blending of cement. *(From an article by H.U. Schalkhauser in ZKG).*

(1) feeding empty paper bags to packing machines

(2) truck and wagon loading, with simultaneous loading of two or more trucks/wagons

(3) feeding, weighing and cutoff of bulk loading of carriers and bulk wagons

(4) billing system.

Plate 7.6.3 Arrangement of cement silo for multiple extraction. *(Brochure of Clandius Peters).*

Layout needs to be designed so that supervision is facilitated.

6.9 The railway siding is for bringing in loaded wagons for coal, slag, gypsum and other materials. It brings in empty wagons for cement and takes away the loaded wagons. There is a lot of movement of rake loads of wagons in and out. Design of railway siding is a specialist's job and should be entrusted to specialists.

6.9.1 Another cardinal principle of layout design is that truck and human traffic does not cross railway siding. This is for reasons of safety and to avoid traffic interruption.

6.10 The layouts in Figs. 7.5.1 and 7.5.2 were developed keeping these factors in mind.

6.11 Various aspects of designing layouts for dispatches of cement have been described in detail, including illustrations, in Chapters 34 to 37 in Section 6 of the author's book **"Handbook for Designing Cement Plants."**

6.12 Plate 7.5.4 in Chapter 5 shows the layout of cement mills and packing and dispatch sections of a 6-mtpa cement plant (split location). It illustrates the various factors to consider when designing layouts for the packing and dispatching facilities mentioned above.

CHAPTER 7

Developments in Existing Machinery and New Machinery Now Available for making Cement

7.1 In the last decade, cement plants have jumped up in size, requiring machines of correspondingly large capacities. Cement machinery manufacturing has also progressed in step with the growth in size of cement plants so that in most cases single machines are capable of delivering the sectional outputs. This is outlined in Table 7.2.1 in Chapter 2 and Table 7.3.1 in Chapter 3.

7.2 Along with the increase in size, the operational efficiency of cement plants has increased, as is evident from reduced fuel consumption, which stands at <700 kcal/kg clinker, and reduced electrical energy consumption, which stands at <75 kWh/ton cement.

This has been brought about by the development of low-pressure cyclones for preheaters and greater use of vertical mills and roller presses (RPs) to grind raw materials as well as clinker and coal.

7.3 As the kiln is now doing mostly sintering, the conventional length/ diameter ratio of 15-16/1 is no longer relevant.

Therefore, for higher capacities, two support short kilns with hydraulic drives have come into vogue.

7.4 In the following paragraphs developments in existing machinery and new machinery are discussed briefly to illustrate the present composition of cement plants.

7.5 Crushers

For large plants of +10,000-tpd capacity, crusher capacity is about 2000 tph.

Since VR mills and RPs are now universally preferred, generally single-stage crushing is adequate. However the primary crusher has

Designing Green Cement Plants
http://dx.doi.org/10.1016/B978-0-12-803420-0.00045-7

to take very large size stones of 1.5 m^3 in size by way of feed from correspondingly large-capacity shovels, with bucket capacities of +13 m^3.

Thus most commonly used crushers are impact/hammer crushers.

Capacities of 2500–3000 tph are available.

Compound impact crushers with two rotors in the same casing reach a capacity of 2000 tph and receive a maximum feed size of 3 m^3 (maximum length ∼1.9 m). See Plate 7.7.1.

7.5.1 Mobile crushers

Mobile crushers to receive stone of 1.5 m^3 at the rate of 2000 tph or so are not available at the moment.

Mobile crushers are mostly jaw crushers. With them two-stage crushing is necessary.

All in all, mobile crushers that appeared to be a good idea for plants of 3000-tpd capacity may not be feasible for very large plants.

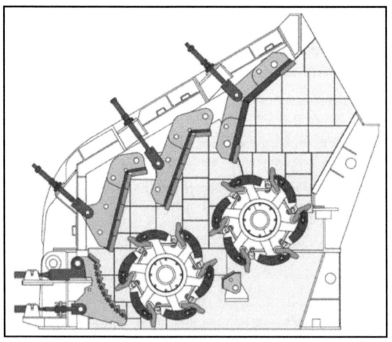

Plate 7.7.1 Hazemag AP-CM compound primary crusher. *(Source: Hazemag Brochure).*

7.6 Earth-moving equipment for mining in large cement plants

For a 10,000-tpd capacity single kiln, stone raised and supplied by quarries is on the order of 2000 tph, based on two shifts working 6 days per week and allowing for overburden to be set aside.

This requires various quarry equipment, as shown in Table 7.7.1.

1. Wagon drills

Even for large-capacity plants, the size of bore holes drilled for blasting is 100-115 mm in diameter.

The number of holes and the depth of drilling depend on the length and width of the face to be opened for excavating stone and feeding it to dumpers for transport. It is entirely feasible to use more than one wagon drill at a time in the same area. Maximum size and capacity defined by length in meters drilled in a given time do not pose any problems.

2. Shovels

It is ideal to have one shovel working on a given face at a time. Assuming one shovel, its capacity has to be about 2000 tph.

Shovels of this capacity are available. The size of bucket ranges from 11 to 15 m^3.

3. Dumpers

Rear dumpers are used to transport excavated stone to the crushing plant. Normally one shovel works with three dumpers. The

Table 7.7.1 Quarry Equipment for a 10,000-tpd Clinkering Unit

1	Wagon drills	2 nos drilling 90-115 mm holes; drilling rate not less than 200 mm/min
2	Shovels	option 1 1600-tph capacity 1 + 1[a] option 2 900-tph capacity 2 + 1[a]; capable of delivering 1500-1800 mm size stone
3	Rear end dumpers	100-tph capacity 7 + 2[a]
4	Bulldozers	2 nos; capacity to be judged by the amount of overburden removed in quarries

Based on sizes and capacities mentioned in brochures of Bharat Earth Movers Ltd.
[a]Is spare.

number of dumpers depends on the distance from quarry face to crushing plant and the time it takes for a return trip.

Dumpers with a heaped capacity of $\sim$80 m^3 are available. Thus dumpers for transporting stone to the crushing plant to match shovel capacity are available.

4. Bulldozers

These are used mainly to clear area and remove overburden and for preparatory work preceding mining operations. They have a great many uses in the plant also. Size and capacity pose no problems.

5. Rippers

The possibility of using rippers to obtain stone instead of drilling and blasting may have to be kept in mind in special cases.

7.6.1 Planning mining operations—green mining

As mentioned in an earlier chapter on green mining, mining operations must plan from the beginning to leave as small a footprint as possible.

7.7 Stacker reclaimer systems for limestone and coal

Stacker reclaimer systems for limestone

These systems serve two purposes. First, they are storage systems for crushed limestone. The quantity to be stored depends on the number of days of stock to be maintained.

Second, they achieve preliminary blending or pre-blending of material while forming and building up stockpiles.

The second function is very important as it complements the blending achieved by the continuous blending and the storage silos installed after the raw mill. The two together achieve the desired blending and keep the standard deviation of raw meal fed to the kiln to $\pm$0.2% or less.

The deposit geological report indicates the total blending to be achieved.

Commonly, using the chevron method to build-up stockpiles, a blending effect of 5-6/1 can be achieved in the pre-blending stage.

In special cases where this needs to be greater, either the build-up method can be modified or the number of layers in the completed pile can be increased to give a greater blending effect.

Stockpiles receive crushed stone directly from the crusher, so the stacking rate must match or be greater than crusher capacity.

In linear stockpiles, one pile is built up to contain stock (in tons or in number of days' requirement). A second pile of similar dimension and capacity is then built up either in same line or alongside.

Choice of arrangements of stockpiles depends on factors like:
- availability of space and the prospects of duplication of the plant capacity.
- if a second crusher is proposed for duplication, the stacker reclaimer system would be duplicated.
- if SR systems to suit plant capacities of 5000-10,000 tpd are available.

7.7.1 Design of stackers and reclaimers

A great many choices are available. Selection should be done in consultation with designers and the suppliers of the SR systems.

7.7.2 Stacker reclaimers for coal

Coal is commonly procured from more than one source. It is often imported. Thus the pre-blending function is as necessary for coal as it is for limestone. The storage systems for coal are now invariably SR systems.

On the one hand, tonnage of coal required is small compared to limestone; on the other, the number of days of stock to be maintained is about three times that for limestone. The quantity of coal to be stored is about 40% that of limestone.

There are two types of stockpiles for coal in common use, circular stockpiles and linear stockpiles.

The capacity of a circular pile cannot be increased, whereas that of linear piles can be increased.

Though quantities to be stacked fewer, dimensions of stockpiles for coal tend to be comparable to those for limestone for two reasons. Firstly, the height of pile is limited to 3 m because of the inherent fire hazard. Secondly, bulk density is half that of crushed limestone.

In the case of a 10,000-tpd plant, end-to-end length of two stockpiles would be $\simeq 800$ m (including coal for CPP), compared to $\simeq 450$ m for limestone.

The layout should be designed taking into account requirements of stockpiles on ground level.

Coal is also required for the captive thermal power plant. It may be sizable. Further, coal received for the power plant is of a different grade (generally inferior) and comes from different sources.

While both come by rail in wagons (or by truckload when collieries are close by) receiving and handling systems for incoming coal have to be designed to cater to both. They would probably be arranged side by side. See Figs. 7.5.1 and 7.5.2 in Chapter 5.

7.8 Vertical roller mills for grinding raw materials and clinker and slag

Presently vertical roller mills (VRMs) have replaced ball and tube mills for grinding both raw materials and clinker in cement plants large and small. RP mills alone or in combination with ball mills for secondary grinding are another preferred choice.

Thus, VRMs or RPs should be available in size and capacity to meet the requirements of large plants.

Table 7.2.1 shows capacities of raw mills and cement mills for cement plants of capacities ranging from 5000 to 10,000 tpd of clinker.

Capacities for cement mills vary widely, depending on the type of cement(s) being made.

For a 10,000-tpd clinker capacity, cement grinding capacities are ~640 tph for OPC, 900 tph for PPC and 1600 tph for BFSC.

7.8.1 In case of raw mills, capacities range from 450 to 900 tph for plant capacities of 5000–10,000 tpd.

Quite a few manufacturers have designs for capacities ranging from 700 to 900 tph.

One designer has come up with multi-drive units, with which connected power can go up to 12,000 kW from three or four drives. Therefore it should be possible to use a single mill even for a plant of 10,000 tpd. See Plate 7.7.2.

7.8.2 The situation is a little difficult for cement mills. Slag is difficult to grind, and if the plant is making slag cement the production capacity required would be ~ 2.5 times that of a plant making OPC only. Thus in case of cement mills it is necessary to use multiple units. This is also necessary for dedicated mills.

Plate 7.7.2 GBR Pfeiffer MVR vertical roller mill. *(Source: Pfeiffer Brochure).*

OPC, PPC, BFSC and slag have different grind abilities and must be ground to different finenesses. The same mill gives very different outputs when producing different products, as the following table shows.

Mill connected power 9000 kW

Material to be ground:	OPC	PPC	Slag
Fineness Blaine	3000	4000	5000
Output in tph	415	430	230

Selection of cement mills has therefore to be done very carefully.

7.8.3 Each designer and manufacturer has its own configuration of components like rollers and tables. All now employ external circuits. Each one has its own design of high-efficiency separator incorporated within the mill.

Thus the size and system of VRMs are not a problem in the case of large cement plants except for requiring multiple units for cement.

7.8.4 VRMs for grinding coal

Capacities required for mills for grinding coal range between 50 and 120 tph for plants of 5000- to 10,000-tpd capacities, depending on coal consumption. The representative figure for fuel efficiency currently is 700 kcal/kg or ~16% consumption.

Coal mills are available from all manufacturers to meet these requirements.

One further option is available in E Mills, that is, ring ball mills.

7.9 Roller press and auxiliaries

A great many options are available when grinding systems with RPs are used.

1. RP in open circuit followed by ball/tube mill for secondary or finish grinding
2. RP in closed circuit followed by ball mill in closed circuit
3. RP alone.

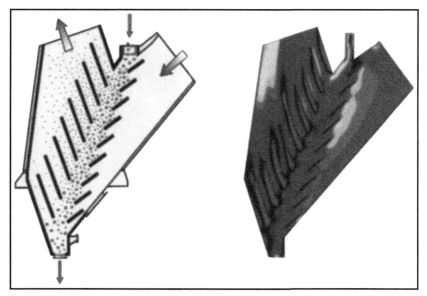

Plate 7.7.3 V separator from Humboldt Wedag. *(Source: Humboldt Wedag Brochure).*

A major component of a grinding circuit with RP is a V separator, or a preliminary separator which separates a coarse fraction before sending the rest to the high-efficiency separator. See Plates 7.7.3 and 7.7.4.

RPs with roller diameters of 2000–2500 mm are available and can be fitted into cement plants of 5000- to 10,000-tpd capacities. The circuits and their components must be designed and selected on a case-by-case basis.

7.10 All in all it can be said that for crushing and grinding operations (in all three departments) machinery large enough in size and capacity is available.

7.11 Continuous blending and storage systems

The continuous blending system is the universally used system for blending ground raw meal before it is fed to the preheater for further processing. As the name suggests, the same silo also serves as the storage silo.

It is the convention to have a 2.5-day stock requirement of the blended raw meal.

Plate 7.7.4 Roller press. *(Source: Humblodt Wedag Brochure).*

On this basis the blending and storage silo/silos should have capacities of 23,000, 34,000, and 46,000 tons, respectively, for cement kilns of 5000-, 7500- and 10,000-tpd capacities.

These capacities need silos of the dimensions shown in Table 7.5.4.

Silos of up to 60,000 tons have been constructed.

Different manufacturers have their own blending system designs. Hence the configuration at the bottom of the silos varies from manufacturer to manufacturer, though all aim to carry out blending simultaneously with the filling process. Incoming raw meal is spread over a large area of the silo by a network of distributing air slides at the top.

Layers so formed are broken as raw meal travels to one or more exit points. In the process blending takes place more or less in the same manner as reclaiming across the cross-section of a stockpile.

If the standard deviation of stone as quarried is 4, in the pre-blending operation it is reduced to 0.8. Raw meal fed to a kiln should have a standard deviation not exceeding 0.2. The blending effect in a continuous blending silo is on the order of 5-6. Raw meal as received is then blended to standard deviation of 0.2, requiring a blending effect of 4, which is feasible.

A prerequisite of the continuous system is the continuous monitoring of raw meal both at inlet and outlet by an X-ray analyzer or similar equipment.

In one variation of the continuous system there is small-quadrant blending system at the bottom to enhance the blending effect.

Another option is to install two blending silos and extract raw meal simultaneously from both of them. A further blending operation thus takes place while it is fed to preheater.

Designers try to minimize air consumption and hence power consumption during a continuous blending operation by sectioning aeration of the bottom.

7.12 Kiln feed systems

In principle, the kiln feed systems remain the same as for plants of a 3000-tpd capacity.

There are two major considerations.

1. Taller preheater heights because of higher capacities require taller bucket elevators.
2. Up to 7500 tpd, one preheater stream is feasible. Beyond that there may be two or more preheater streams, either the same size or in proportion to the division of raw meal between kiln and calciner.
3. When there is more than one stream, raw meal withdrawn from the blending and storage silo needs to proportionately distributed between two or more kiln feed systems.

    Thus there are as many kiln feed systems as preheater streams.

    A standby must also be provided for uninterrupted operation.

4. In low NO_x calciners, both raw meal and fuel are fed at two levels. There are a great many further developments in the design of calciners, such as down draught and secondary fluid beds.

5. Division of feed at two points in a calciner may be obtained by a dividing or regulatory damper. To this extent, the design of kiln feed systems is more demanding in large plants.
6. Large-capacity six-stage preheaters require tall preheater towers. Raw meal is fed between the first and second stages, requiring tall elevators. These kiln feed elevators must have 100% availability and standby.

7.13 Waste heat recovery systems

Power stations using waste heat from kiln and cooler gases have become commonplace. This is one of the distinct features of a green cement plant.

There are many systems available: simple Rankine cycle. organic Rankine cycle, Kalina cycle, and so on. They have been discussed in Section 5.

If the heat content in exhaust gases at the end of sixth stage is not sufficient, it is sometimes increased by tapping gases from two stages; that is, the preheater is used either as a five-stage or six-stage preheater, requiring the ability to divide kiln feed as desired between the sixth and fifth stages and the fifth and fourth stages. See Flow chart 5.6.6.

7.14 Pyroprocessing section

7.14.1 Rotary kiln

The significant development in kilns of large capacity is two support kilns.

The length/diameter ratio is now from 11 to 13/1 instead of the 15-16/1 of yesteryear. Because of reduced specific fuel consumption, which stands between 700 and 650 kcal/kg for six-stage preheater kilns, the specific output of dry process calciner kilns has now increased to about 4.9 tpd/m^3 for three support kilns. It has increased further for two support kilns to about 5.42 tpd/m^3. See Table 7.7.2.

There are significant developments in the mechanical design of kilns and their drives, such as:

1. Riding rings are of splined design.
2. The drive can be through the roller at the inlet end.

Table 7.7.2 Two and Three Support Kilns

Plant Capacity with Calciner (tpd)	Two Support			Three Support		
	Size Dia (m)	Length (m) l/d 12	Specific Output (tpd/m³)	Size Dia (m) l/d-	Length (m) Avg 15.62	Specific Output (tpd/m³)
4000	4.55	54	5.6	4.4	67	4.9
5000	5	60	5.1	4.75	74	4.7
6000	5.25	62	5.5	5	78	4.7
7000	5.5	66	5.4	5.25	82	4.8
8000	5.75	69	5.4			
8500				5.5	87	5
9000	6	72	5.5			
10,000				5.75	91	5.1
11,000				6	95	5.1
	Avg		5.42			4.9

Source: FLS Brochure.

It will be dual in most cases (planetary gear; variable speed motor/hydraulic)

3. Supporting roller and bearings assemblies are of self-aligning type.
4. Hydraulic thrust roller monitors movement of the kiln.

The kiln hood is modified to install a Gunnax system to inject whole tires into the burning zone of the kiln. See Plates 7.7.5–7.7.7.

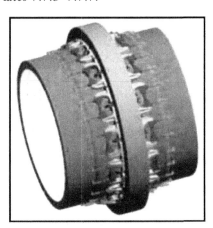

Plate 7.7.5 Tangential tire suspension system. *(Source: FLS review).*

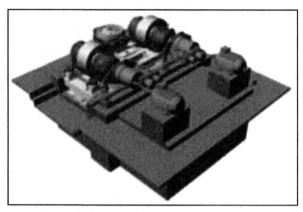

Plate 7.7.6 Friction drive of kiln through supporting rollers. *(Source: FLS review).*

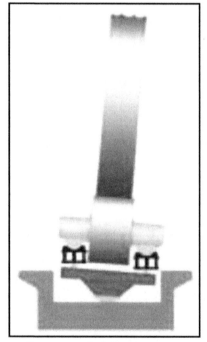

Plate 7.7.7 Self-aligning roller support for kiln. *(Source: FLS review).*

7.14.2 Fuel firing systems

In precalciner kilns fuel is fired in the kiln and calciner generally in the ratio 40:60.

When coal is the fuel, it is customary to install a coal mill at the preheater end. Two metering systems withdraw pulverized coal from the fine coal hopper and feed it pneumatically to the kiln and calciner.

Coal burners at the kiln end are now multi-channel burners designed to receive and fire more than one fuel, coal and oil or coal and gas. They are also designed to fire alternative fuels which can be of different types, including pulverized solids, sludge, oil, biomass, etc. See Plate 7.7.8.

Fuel firing in the kiln at the discharge end has to be designed carefully considering the possibility of alternative fuels.

The Hot Disc system is designed to fire whole/shredded tires and other solid fuels at the kiln inlet and/or in the calciner. Both Gunnax and Hot Disc are FL Smidth Pvt. Ltd. (FLS) products.

See Plates 7.7.9–7.7.11.

7.14.3 Alternative fuels

Most new cement plants use alternative fuels/carbon neutral fuels to reduce emission of greenhouse gases (GHG). This aspect has already been discussed in Section 4. The extent of alternative fuel fired is 20-30%. When there are two or more preheater streams and calciners, it may be more convenient to fire the calciner in one stream with AF and the other with regular fuel. The feasibility of this arrangement from the point of view of quality control and other considerations should be considered.

Plate 7.7.8 Duoflex multi-channel multi-fuel burner. *(Source: FLS Brochure).*

Plate 7.7.9 Hot Disc for feeding whole tires and other solids. *(Source: FLS Brochure).*

For example:

Division of fuel between calciner and kiln = 60:40

If kiln has two streams of preheater and one calciner

Raw meal will be divided between two streams in accordance with their sizes

Calciner will receive 60% fuel

Plate 7.7.10 Gunnax system for firing whole tires. *(Source: FLS review 148. Environmental constraints).*

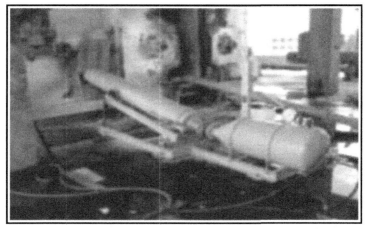

Plate 7.7.11 Gunnax firing waste canister. *(Source: FLS review 148. Environmental constraints).*

If AF is 20% of total, then division of fuel would be 40% regular and 20% AF

Calciner has thus to receive two kinds of fuel from two fuel firing systems

As mentioned above, whole tires can be introduced into the kiln as fuel either from the kiln hood or the inlet end.

Shredded tires and other shredded solid fuels can be introduced at the kiln inlet end or in the calciner.

7.14.4 Firing fuel in calciner

A calciner receives about 60% of the total fuel. As mentioned above, part of this fuel can be alternative fuel.

Many kinds of alternative fuels are available. At the design stage it must be ascertained what type(s) of AF is continuously available in what quantities and the consistency of calorific value and other properties.

A fuel firing system then has to be designed to fire:

Main fuel + AF in calciner in predetermined proportions

Main fuel in calciner, AF in kiln.

On occasion, calciners may have to receive two kinds of alternative fuels, sometimes even simultaneously.

A suitable system of storing, extracting from bins and metering has to be designed.

See Figs. 4.6.1–4.6.11 in Section 4 for arrangements for firing fuel in kiln and calciner.

7.14.5 The properties of AF

As mentioned earlier, fuel is fired in the calciner at two or more levels to control atmosphere within the calciner (oxidizing/reducing to control NO_x formation). Therefore there have to be multiple feeding points for regular fuel and for AF.

The control system would include gas analyzers which in addition to normal components would analyze gas for NO_x, SO_3, Cl_2, and heavy metals.

7.15 Preheaters

Most common preheater systems are six-stage preheaters because they reduce fuel consumption.

The geometry of cyclones is designed to have a low-pressure drop on the order of ~450 mm even in six-stage preheaters. Total pressure drop including the calciner is around 550 mm.

There is no horizontal shelf in the volute of the incoming duct. Each manufacturer has developed designs for dip tubes that ensure high efficiency and low-pressure drop at the same time.

A significant development is the maximum size of cyclones in a single stream. Single streams are now possible for kiln capacities up to 6000 tpd (cyclone diameter ~9-10 m). Beyond this, two streams going up to 11,000 tpd are installed.

Table 7.7.3 shows recommendations of FLS regarding plant capacity and the number of streams.

Because of the increase in capacities of a single kiln above +10,000 tpd, two- or three-stream preheaters with one or two calciners have been installed.

Fig. 7.7.1 shows such an arrangement,

The main reason for low-pressure drops in cyclones is the low velocities at the inlet. These are lower than 15-16 m/s.

In spite of such low inlet velocities, efficiencies as high as +95% are claimed for top cyclones on account of the innovative designs of dip tubes.

Table 7.7.3 Kiln Capacity and Number of Preheater Streams

Capacity (tpd)	Preheater Streams	FLS System[a]	Kiln Support	Calciners
4000	1	sp	3	1
5000	1	ilc-e	3	1
6000	1	ilc	2/3	1
7000	2	ilc/slc	2/3	1
8000	2	ilc, slc, slc-I, slc-d	2/3	1
9000	2	ilc, slc, slc-I, slc-d	2/3	1
10,000	2	ilc, slc, slc-I, slc-d	3	1
11,000	2	ilc, slc-I, slc-d		1
	3	slc-d	3	2

[a]Nomenclature from FLS Brochure dry process kiln systems.
Source: above-mentioned FLS Brochure.

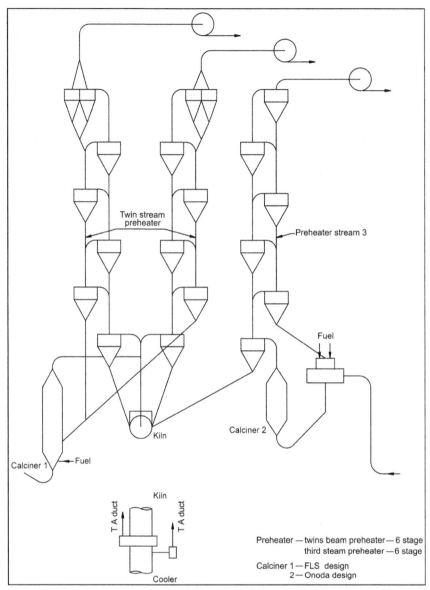

Figure 7.7.1 Flow chart for kiln with three streams of preheater and two calciners (for illustration only).

As mentioned earlier, a six-stage preheater may also be used as a five-stage preheater to use the higher heat content of exhaust gases to generate power in WHR systems.

7.15.1 It must be mentioned here that one prerequisite of high-efficiency low-pressure drop cyclones is that they must operate close to design/full load capacity. It is not possible to operate them at partial capacity.

Therefore it may be prudent to have two streams instead of one so that the plant can be run at least at 50% capacity in emergencies.

7.16 Calciners

There are any number of designs of calciners and perhaps as many ways in which they can be installed in the circuit.

The main variations are:

in line and offline calciners

fluid bed calciners

low NO_x calciners

downdraft calciners

secondary calciners attached to preheater cyclones for two-stage calcining.

The ONODA RSP calciner is the forerunner of the downdraft calciner that is so common today.

Various types of calciners have been shown in Figs. 9.2–9.10 in Section 2, Chapter 9 of the author's book, "Handbook for Designing Cement Plants."

The location and number of feed points for raw meal and coal/oil/AF also vary from design to design although the objective is the same: to complete the combustion of the injected fuel and to maintain the temperature profile that will ensure calcination.

Fig. 9.10 in the author's book shows arrangements of a DD Furnace that was designed to reduce NO_x.

It is further developed to introduce both raw meal and fuel at multiple levels. A reducing agent containing 5% ammonia is introduced in the reduction zone to bring the NO_x down further. See Fig. 7.7.2 and Plate 7.7.12.

Fuel is also fired from the top in some designs like the Onoda RSP calciners in swirl chambers and in downdraft calciners.

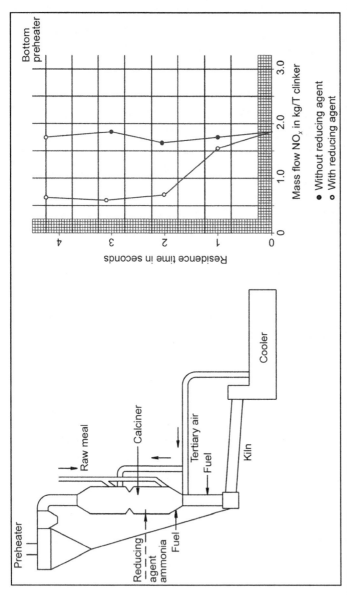

Figure 7.7.2 Control of NO$_x$ by multiple feed of fuel and raw meal and injection (reducing agent ammonia). (*Source: VDZ activity report 2003-2005*).

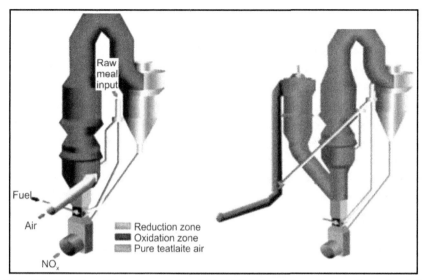

Plate 7.7.12 (a) Low NO_x calciner ILC and (b) SLC D NO_x calciner. *(Source: FLS review 148. Environmental constraints).*

Calciners with longer retention times like fluid bed calciners can take coarsely ground fuels. If the plant is large enough, two coal mills can grind coal to different finenesses and supply them to kiln and calciner respectively.

Arrangements for two-stage calcining are shown in Figs. 9.7 and 9.8 in Section 2 of the above-mentioned handbook.

For large-capacity plants the three-stream preheater and two calciner logical solution is shown in Fig. 7.7.1.

AFs used are often in the form of granules (rice husk), shredded material (shredded tires and biomass) and briquettes (bio mass, municipal solid wastes).

Since the calciner receives 60% of the total fuel, often it is the calciner in which AF is fired.

If it is planned to use AF from the beginning, the design of the calciner can be modified to receive the selected AF. The designer can also help with design of the fuel feed and firing system in the calciner.

When AF is introduced in a running plant, the advice of the preheater and calciner designers should be taken when designing the AF firing system to suit specific AFs.

7.17 Kiln bypass systems

Old preheater kilns faced problems when using raw meal containing high chlorides (and alkalis). A solution was found by bypassing the quantity of gases leaving the kiln so that the concentration of alkalis and chlorides did not reach levels such that the coatings formed in the bottom two stages seriously hampered running of the kiln.

With the advent of calciners, fuel and gas flow were divided between kiln and calciners, thereby reducing greatly the quantity of gases leaving the kiln. This reduced the need for bypass at the kiln inlet end.

However the increasing use of AFs of a great variety coming from a great number of sources sometimes necessitates installation of the bypass to get rid of not just the chlorides but also heavy metals like. Therefore in many plants using AF a kiln bypass is installed.

The system of bypass is similar: the collected gases are cooled and passed through filters. They may not be led back to the system. See Fig. 7.7.3.

They may also be taken directly to a WHR system to augment the heat content of waste gases after the preheater, when there are no undesirable elements and bypass *per se* is not required. See Fig. 5.6.12.

A bypass system increases fuel consumption and hence the quantity of bypass is kept as small as possible.

7.18 Clinker coolers

There have been a great many developments in the design of clinker coolers. In addition to the Cross Bar, Coolax and Pendulum coolers mentioned earlier there are new designs under the trade names of Multimovable and SF Cross Bar Coolers by FLS, Pyrofloor Coolers by KHD, Repol and Poly-track Coolers by Polysius and N Cooler by Claudius Peters, and Pendulum cooler by IKN, the leading designers of clinker coolers.

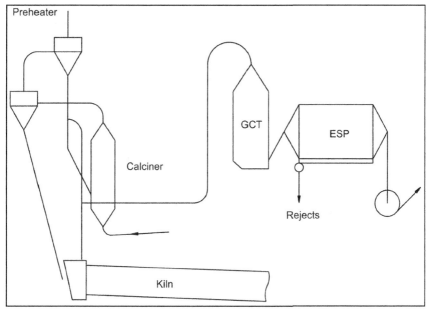

Figure 7.7.3 Flow chart for bypass arrangement at kiln inlet to remove objectionable elements.

7.18.1 Though they differ considerably in detail from one another, there are some similar features like:

1. a steeply inclined static grate for the first 8–10 rows
2. many designs have a walking floor
3. individual control of flow of cooling air to grate plates; quantum of cooling air is on the order of 2–2.2 nm^3/ kg clinker
4. specific loading in terms of tpd/m^2 grate area is about 40–45
5. there is no spillage in almost all designs, eliminating spillage drag chain and head room required for it
6. single width coolers of widths up to 7 m and total grate area of 290 m^2 for kiln capacities of 12,000 tpd
7. not much is known about air loading in terms of nm^3/ min/m^2 along the length of the cooler
8. almost all designs have roll crusher either midway or at the end to break clinker lumps

9. cooler drive is hydraulic in many cases

10. when there is spillage, it is often transported pneumatically

11. all designs claim modular construction for speedy assembly and expansion when required.

7.18.2

1. in cross bar coolers, grate plates do not move; a system of cross bars above them push clinker

2. in coolers with a walking floor, the width has a number of plates. While all plates move forward together, they return at different speeds and the sequence of return is also different along the length. See Plate 7.7.13.

Also see Plate 7.7.14 showing a poly-track cooler of Polysius.

7.18.3 Operational efficiency of new design coolers is high at +70%. Power consumption for the drive and fans has come down to about 3-5 kWh/ton clinker.

7.19 Clinker storage

Clinker is conveyed from clinker coolers by deep bucket conveyors to clinker storage. In most cases clinker is stored in large-diameter conical shades with a short cylindrical height at the base.

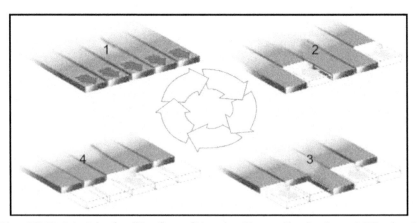

Plate 7.7.13 Principle of walking floor of N cooler. *(Source: Claudius Peters AG Brochure).*

Plate 7.7.14 Polytrack clinker cooler. *(Source: Polysius Brochure).*

A variation of this scheme is to have part of storage below ground level. See Fig. 7.7.4.

It can also be stored in silos. Silos are particularly suitable when clinker is dispatched for export or for grinding at a separate location.

A standby is a must for a clinker conveying system, as is interchangeability when there are two (or more) production lines.

Pan-conveyors are suitable for conveying clinker from clinker storage to cement mills, particularly when clinker is discharged at multiple points.

7.20 Cement mills

Types of cement mills have been discussed in paragraphs 7.8 and 7.9.

Most plants make at least two types of cement. By definition, a green cement plant must make blended and or composite cements. This creates many possibilities in design.

1. mills will be more than one to make OPC and blended cements like PPC, BFSC
2. mills will be identical in type and size or can be different
3. mills will be dedicated to grinding specific types of cements
4. depending on the size of the plant there will be multiple mills to make same type of cement.

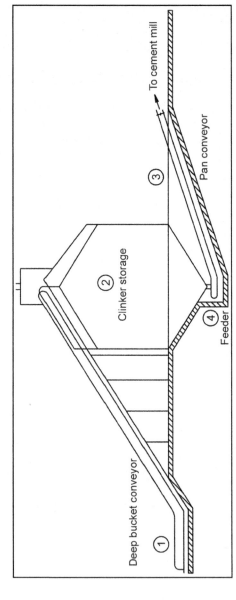

Figure 7.7.4 Alternative clinker storage system for large cement plants.

All these possibilities require careful planning of the cement mill layout at the design stage.

See Table 7.3.1 in Chapter 3.

7.21 Cement storage

Chapter 6 describes the various options that need to be considered when designing cement storage and dispatch.

Cement storage silos are similar to the blending silos for raw meal except that they are primarily used for storage.

Bottom arrangement is similar to the continuous blending silo with a central inverted cone or a number of smaller cones and outlets.

Cement storage silos are often designed to receive more than one type of cement. Cements so stored can be withdrawn for packing/bulk dispatches simultaneously.

A common construction is a ring silo within the larger outside silo. This arrangement can hold two types of cement. The outer ring silo can be further compartmentalized. See Plates 7.6.1 and 7.6.2 in Chapter 6.

In large plants making two or more types of cement it may be worthwhile to have:

1. silos earmarked for specific types of cement
2. silos earmarked for dispatches of bagged and bulk cement
3. configuration of silos depending on the proportion of dispatches by rail and road.

In most cases, dispatches must be carried out in bulk and bagged by rail and by road simultaneously for all types of cement the plant is producing.

7.21.1 Cement packing and bulk dispatches

With the increase in size of cement plants, volumes of dispatches and rates of packing of bagged cement must increase correspondingly. Bagging machines with capacities of +220 tph are now available. Almost exclusively rotary packing machines with electronic weighing devices are installed.

There is corresponding development of automatic feeding of bags (paper) to the packing machines.

7.21.2 Down the line, bag loading, truck or wagon loading, and palleting machines for loading in containers are also available in capacities required.

Entire bagging and loading operations are remote controlled.

Bulk loading has also been automated. Rates of loading have to keep pace with the volumes handled.

In the case of dispatches by train, bagged or in bulk, restriction of the number of hours available also has to be taken into account.

To design a layout for a packing/cement dispatch section is therefore a challenging job and needs to be taken up as such. See Plate 7.6.3 in Chapter 6.

7.22 Transport of cement and emission of greenhouse gases

Transport by rail (whether bagged or in bulk) will help reduce greenhouse gas emissions, though at present its share is very small.

7.23 General plant layouts

Section 5 deals exhaustively with the development of the layout of large green cement plants.

Layouts in Figs. 7.5.1 and 7.5.2 have been developed for illustrative purposes, taking into account the various factors mentioned above.

7.24 Automation, instrument, and process control

Automation has been an integral part of cement plants for a long time. Presently, with plant sizes rising up to more than 3 million tons per annum in one location, operation of the plant without automation is unthinkable. Today automation encompasses all activities directly or indirectly connected with the manufacture of cement.

Basically automation means process and operational control.

It is therefore primarily concerned with:

1. continuous quality control and its monitoring
2. continuous process control
3. continuous monitoring and control of environment.

To these three are now added

1. control of cement dispatching operations
2. financial aspects like inventory control, cash flow, and so on.

There is hardly any area of cement plant operation that is not touched or that is not dependent on computerization and automation for efficient operation.

7.24.1 Quality control

Quality control encompasses following activities:
1. collecting samples of different materials from different locations
2. transporting them to the laboratory and preparation of test samples
3. actual testing using X–ray analyzers or similar quick testing alternatives, obtaining instant results
4. using results to continuously monitor various operating parameters.

All these are interlinked and must work in perfect unison to achieve the objective of the desired quality at optimum cost and operational efficiency.

Expertise and equipment for automation in quality control has been available for a long time. Total automation was previously optional, however.

With increased sizes of plants, total automation in this area has become a must. Major cement machinery manufacturers like FLS and Polysius (Polysius Knepp Thyssen) have their own systems.

Companies like Siemens, ABB, etc. have specialized instrumentation and control.

7.24.2 Process control

Process control involves measurements of all parameters that have a bearing on (a) quality of product and (b) efficiency in production in terms of consumption of materials and energy (thermal and electrical, and costs of the two).

Measurements have to be continuous and the system must interpret them and give appropriate signals online for correction as required.

In all control systems, heat balances and analysis of electrical power consumption are instantly available. From them, factors for deviation can be readily deduced. These are used for correction. Useful tools like material and heat balances and energy consumption are devised to locate deviations and their sources and are used for continuous corrections, using systems like Fuzzy Logic.

Trend curves are used for long-term control.

A new concept is dashboard control. Like the dashboard of a car or cockpit of an airplane, necessary operational parameters are presented on a screen which shows deviations if any and alerts top management to the need for corrective actions.

7.24.3 Environmental control

This refers to particulate emissions from various vent stacks in the cement plant and noise levels.

It is statutorily binding to comply with standards for particulate emissions as laid down by the concerned pollution control board of each country.

Levels beyond which noise levels of a production department should not go have also been defined by most PCBs. Cement plants try to remain within those norms.

7.24.4 Greenhouse gas emissions

A new dimension has been added to environmental control which is basic to all green cement plants. This is emission of GHG. It behooves makers of green cement to monitor emissions of CO_2 and other harmful elements and compounds.

Cement plants must therefore install necessary equipment and monitor GHG emissions regularly.

7.24.5 Advent of Internet and mobile phones

With advances in PCs (personal computers) and the advent of the Internet, it has become possible to make available all data to all concerned executives of the company whether in the plant or elsewhere at any time.

Laptops become movable monitors in the hands of executives responsible for process and quality control in cement companies (CC).

Mobile phones further enhance the mobility of processed data. It is now available at the touch of a finger anywhere anytime and instructions can be passed on instantaneously.

7.25 Key performance indicators (KPIs)

The concept of sustainable development (SD) has shifted the emphasis to growth along with conservation of resources and protection of the environment.

The World Business Council for Sustainable Development (WBCSD) commissioned Batelle to study SD in the cement industry. Batelle published excellent reports in 2002 which serve as guidelines for the cement industry.

7.25.1 One report concerned establishing key performance indicators (KPIs). Batelle recommended principal KPIs pertinent to the cement industry. In consultation with the cement industry they also suggested base values.. These KPIs became benchmarks for assessment of performance of existing cement plants and for design of new cement plants.

7.25.2 Cement companies have begun to develop their own KPIs and to include them in mobile/Internet-assisted control systems.

The KPIs have thus become visible. This is the new mobile dashboard control system.

7.25.3 KPIs like energy efficiency, GHG emissions and costs of production (to name a few) are now available to the concerned executives on mobile phones and on laptops, like operational parameters presented on dashboards of cars or in cockpits of airplanes.

It is for the CC to decide on the depth and details to be shown in each section of the dashboard.

This concept is fast catching up as it is convenient and available anytime anywhere.

7.26 Automation systems of different manufacturers

All cement process and machinery designers have their own well-developed and proven automation systems for process and quality control and these are well known by their brand names.

Companies like Siemens and ABB who are specialists in electrical power and instrumentation systems also have well known automation and process and quality control packages specially developed for cement plants.

Expert advice should be taken before selecting a system or systems that will best suit a given project, its raw materials, the fuel and the quality parameters of cements to be produced.

It is logical that the cement machinery manufacturers selected should also engineer and supply quality and process control systems if this falls in their range of products.

7.27 Motors and gear boxes

Large mills require motors and gearboxes with high ratings, from 3000 to 6000 or higher kW ratings.

Large motors for ball/tube mills can be ring motors that will eliminate the gearbox and gear drive.

Vertical mills require special gearboxes that have horizontal input and vertical output shafts and also have to bear thrust loads.

As mentioned earlier, VRMs with multi-drive units have been developed (total drive rating of 12,000 kW is obtained by four units of 3000 kW).

The number of units may sometimes be dictated by the largest rating of motor and gearbox available.

In kilns, one of the roller stations provides the drive to the kiln rather than through the motor-gearbox-gear/pinion chain.

Hydraulic variable speed drives are coming into vogue for kiln and cooler drives.

7.28 Electrical switch gear

A power distribution system is a vital component in a cement plant, beginning with the substation of the grid in the plant and ending with power supply to individual drives.

There should not be any problem in this respect in the case of large cement plants of +10,000 tpd capacity.

7.29 Integration of power systems

Today's large cement plants have three sources of power.

1. grid power (main source of power)
2. captive thermal plant TPP (mainly fired by fossil fuel)
3. power generated by WHR system.

The plant has to make judicious use of these to ensure continuous availability of power in the required quantity year round at optimum cost.

It may be profitable to sell WHR power to the grid, or to buy less grid power and depend on the plant's own TPP and WHR power.

Availability of power from each source year round must be carefully evaluated and a careful plan prepared to share power among these sources so that the net power bill is minimal and power is continuously available.

7.29.1 A new dimension is added if the cement company decides to invest in renewable sources of energy like wind or solar power.

Sources of wind or solar power may be at a different location. Even then power generated can be "wheeled" to the cement plant.

These are problems specific to large green cement plants of today that require solutions.

CHAPTER 8

Operational Efficiencies of Modern Large Cement Plants

8.1 The foregoing paragraphs have given a sufficient idea of the major machinery and auxiliaries in today's large cement plants.

These would be:

1. single-stage crusher
2. VRM or roller press system for raw material grinding
3. continuous blending systems
4. six-stage preheater and calciner
5. kiln and cooler of latest design
6. VRM for grinding coal
7. VRM or roller press system for making cement
8. machinery for bulk handling and dispatch of cement
9. materials handling equipment that consumes low power
10. Dust-collecting equipment that is efficient and yet consumes less power.

8.2 Specific heat consumption

Presently specific heat consumption ranging between 700 and 650 kcal/kg is common.

An amount of 700 kcal/kg can be taken as a representative figure.

8.3 Specific power consumption

Unlike specific fuel consumption, there is wider variation in specific power consumption on account of:

1. variations in the grindability of raw materials
2. the types of cements made (OPC, PPC and BFSC) and the fineness to which they are be ground, according to proportions of the blending elements.

Even then, specific power consumption varies between 65 to 80 kwh/ton. An amount of 75 kwh/ton can be taken as a representative figure.

8.4 Particulate and other emissions

Permissible particulate and other emissions vary within a small margin from country to country, as per the standards laid down by concerned pollution control boards.

It is now more or less mandatory to monitor these continuously.

An amount of 50 mgm/nm^3 is the commonly accepted norm for particulate emission.

8.5 Monitoring CO_2

Most cement companies the world over are committed to minimizing GHG emissions.

Quite a few companies are gearing up to monitor CO_2 emissions. The amount depends on the types of cements made, the cement/clinker ratio, alternative fuels and their proportion, and also whether a waste heat recovery system has been installed.

This aspect has been dealt with in detail in the previous chapters.

8.6 Manhour productivity

Another yardstick to measure performance of a cement plant is manhours per ton of cement.

There has been tremendous reduction in manhours per ton of cement on account of:

1. Scale of operations
2. Automatic sampling
3. Process automation and control
4. Higher proportion of bulk handling and dispatches of cements
5. Automatic truck- and wagon-loading machines for bagged cements.

8.6.1 A decade ago, for a 1-mtpa plant making OPC, 0.67 manhours per ton was a typical measure of man-hour productivity (see author's **"Handbook for Designing Cement Plants"**).

Presently, with the representative plant capacity rising to that of a 10,000-tpd clinkering unit and 5-mtpa capacity in cement, man-hour productivity has increased correspondingly.

Taking the case of the 10,000-tpd kiln/6.5-mtpa cement selected for developing layouts in Chapter 5 (Annexure 1), the estimated manpower varies between 500-600.

Variation arises because of different proportions of bulk and bagged cements and proportions of dispatches sent by road and rail from place to place.

The extent of mechanization and automation also influences the amount of manpower employed.

Cement companies may outsource maintenance and repairs to some extent.

Local conditions for recruitment of semi-skilled and unskilled labor and the use of contract labor also have an impact on manhour productivity.

Typical manhour productivity lies between 0.18 and 0.13 manhours per ton of cement: a substantial increase!

RECOMMENDED READING

1 Manual on Thermal Energy Efficiency in Cement Industry
2 Manual on Best Practices in Cement Industry
3 Manual on Best Practices in Indian & International Cement Plants

All above by Confederation of Indian Industries Sohrabji Green Business Centre

Capital Costs and Costs of Production

Contents

List of Tables

CHAPTER 1

Capital Costs and Costs of Production

1.1. Structure of capital costs (CC)

The procedure for working out capital costs for a green field cement plant Project and presenting them in a format acceptable to financial institutions is presented in the author's **"Handbook for Designing Cement Plants."**

The procedure remains the same currently.

1.1.1. Naturally updating of costs is required to reflect the size of the plant, which has trebled since that book was written, and also to reflect present-day costs of plant and machinery, electrical needs, civil construction costs, erection and commissioning costs, as well as interest charges.

1.2. New elements of costs

New features of today's green cement plants like blended cements, alternative fuels (AF),waste heat recovery systems (WHRS), and renewable energy (RE) need to be reflected in the calculation of the capital costs of today's cement plants.

1.3. Capital costs

Capital costs depend on the contents of the plant, which have several aspects.

1. Technology and process selected.
2. Choice of machinery. For the same operation there are several options.
3. Margins chosen when sizing major machinery units and auxiliaries.
4. Whether multiple units are preferred to single units.
5. The layout and provision allowed for likely expansion.
6. Available infrastructure. Some countries have to import machinery, motors gearboxes, process control equipment, and the like.

Designing Green Cement Plants
http://dx.doi.org/10.1016/B978-0-12-803420-0.00047-0

More advanced countries have much less import content. Cost of imports includes items like exchange rates and import duties, which push up costs.

1.3.1. All in all it is a subjective exercise as calculation is generally project specific. No two projects are identical, though they may have similarities.

1.3.2. For this reason, over the years, the cement industry has been using yardstick measurement to see if the capital costs for a planned project fall in the range prevailing at any given time. The deviations if any can then be evaluated and accounted for.

The yardstick measurement is the **cost in Rs/annual capacity in tons of cement**. In earlier days cement meant OPC. Now of course there are blended cements and each type has its own yardstick.

Currently capital costs for a green-field-site large cement plant range between Rs. 4000 and 4500 ($73 to $82)/OPC annual capacity.

It may be necessary to increase this by 5% when railway siding is necessary.

1.3.3. Typical capital costs for a 10,000-tpd iln making ~3.5 mtpa OPC and 8.7 mtpa BFSC have been furnished in Table 8.1.1.

This shows the impact on capital costs expressed in **Rs./ton** of annual capacity from the use of blended cements (BFSC).

While capital costs increase from **Rs. 14,000 to Rs. 21,000** million ($255 to $380 million), costs expressed as Rs./ton of annual capacity decrease **from Rs. 4000 to Rs. 2400** ($73 to $43.6) per ton because of the increase in production from 3.5 mtpa to 8.7 mtpa.

1.3.4. The impact of installing an AF and WHR system on the capital costs of the same kiln is shown in the same statement.

1.3.5. It is assumed that the plant installs a system for firing alternative fuels to the extent of 20%. It will require an investment of Rs. 100 (~$1.8) million and also a WHR system based on the organic Rankine cycle.

Table 8.1.1 Impact of Making Green Cement on Capital Costs

Item	Unit	OPC	BFSC
Kiln capacity	tpd	10,000	10,000
Cement production	mtpa	3.5	8.7
		(100% OPC)	(100 % BFSC)
Capital costs	million Rs.	14,000	21,000
	million $	254.5	382
Capital costs	Rs./ton	**4000**	**2400**
	$/ton	72.7	43.6
Capital costs AF	million Rs.	100	100
	million $	1.82	1.82
WHR	million Rs.	1340	1340
	million $	24.4	24.4
Total with AF & WHR	million Rs.	15,440	22,440
	million $	281	408
Capital costs With AF & WHR	Rs./ton	4411 **4400**	2579 **2580**
	$/ton	**80**	**47**
Increase on account of AF & WHR	%	**10**	**7**

Note: above capital costs are without railway siding.
Rate of exchange: 1 $ = 55 Rs.
tpd = tons per day.
mtpa = million tons per annum.

For a plant of this capacity the rating of WHRS is ~12 MW and the investment required about Rs. 1340 ($24.4) million, at $2.0 million/MW as mentioned in Chapter 8 of Section 5.

Taking these two additional investments into account, total capital costs for the two alternatives would increase to **Rs. 15,440** (~$281) and **Rs. 22,440** (~$408) million respectively.

Corresponding costs in Rs./$ per ton would be Rs. 4400 (~$80) and Rs. 2580 (~$47) respectively.

While the impact when making OPC is about 10%, that when making BFSC is about 7%.

1.4. Consultants

Most cement companies wishing to embark upon projects of this magnitude are well equipped to perform techno-economical feasibility studies and consequently select the plant and machinery themselves.

However consultants can play a major role in assisting and guiding cement entrepreneurs to make the right decisions speedily.

This is particularly true when there are new factors to consider, such as alternative fuels (AF) and waste heat recovery systems (WHRS) as well as renewable energy (RE) like wind and solar power.

1.5. Costs of production (CoP)

The impact of present day costs of raw materials and fuel and the use of a higher quantum of blending materials like fly ash and slag as well as the use of wastes for fuel should be reflected in the present costs of production.

Typical costs of production of naked cement ex works were worked out based on the present costs of raw material, fuel, power, etc.

1.5.1. An exercise was then performed to find the impact of substituting fossil fuel by AF to the extent of 20% and by partial use of power generated by a WHRS (Waste Heat Recovery System).

This exercise was done for OPC and also for PPC and BFSC.

The impact of making green cement (blended cements with AF and WHR) has been shown in Table 8.1.2 using naked cement.

It shows the costs of production of OPC, PPC and BSFC without and with AF and WHR.

1.5.2. The impact can be clearly seen.

On the one hand, the CoP for OPC is Rs. 1830/ton (~$33.3)

On the other, the CoP for BFSC with AF and WHR is reduced to Rs. 1420/ton. (~$25.8), a reduction of 24%.

1.6. Bulk dispatches of cement

Overall costs come down further when cement is dispatched in bulk by road, rail or ship.

If the bagging costs are Rs. 160/ton (~$2.9), bulk loading costs will be around Rs. 50/ton (~$0.9).

Table 8.1.2 Impact of Making Green Cement on Costs of Production of Naked Cement

		OPC		PPC 30% Fly ash		BFSC 60% Slag	
		Normal	With AF& WHR	Normal	With AF& WHR	Normal	With AF& WHR
Sr. No.	Item	Costs in rupees per ton cement					
1	Total variable costs	1681	1555	1446	1357	1367	1332
2	Fixed costs At 85% production	146	154	152	123	87	91
3	Total costs of production	1827	1709	1598	1480	1454	1423
	Rounded to	**1830**	**1710**	**1600**	**1480**	**1455**	**1420**
	Relative costs	100	93	87	81	80	78

Thus if 50% of the cement is dispatched in bulk instead of bagged, savings in cost per ton would be Rs. 55/ton (~$1).

1.7. Examples of capital costs and costs of production are based on current cost structures prevailing in India. Naturally there are differences from country to country.

Further, the basis of working itself can be different as there are differences from country to country regarding import content and interest rates on loans. The examples are presented mainly to show the gains obtainable by making green cements.

The results prove beyond a doubt the financial advantages of going green. The extent is subject to variation from case to case.

Note:

1. Capital costs are without railway siding.
2. It is assumed that 1$ = 55 Rs.

SECTION 9

Cement Substitutes

Contents

CHAPTER 1

Cement Substitutes: A Peek into the Future

1.1. As a result of the shifting emphasis toward sustainable development and reducing GHG emissions, research to find substitutes for known varieties of cements has accelerated.

1.2. The limitations of reducing CO_2 by making blended cements will prevail as long as the basic raw material to make cement remains limestone (calcium carbonate). The cement industry has almost reached the limits of fuel efficiency. There is very little scope to reduce it further unless temperatures at which clinker is formed (1400-1500 °C) are lowered substantially.

1.3. Research to find cement substitutes has been therefore directed toward overcoming these limitations. While a great many research institutions are engaged in this pioneering effort, already there are a few which have advanced considerably in their work and show great promise for the future. Among the more promising are:

1 Calera process
2 Novacem
3 Calix
4 Geopolymer cements
5 Aether

All the above are basically green cements.

1.4. At the moment they are in the experimental (pilot plant) stage. It will be some time before they prove their commercial viability in competition with today's cement industry and are capable of producing on a comparable scale. However, the cement industry should take note of these developments and adapt to meet the challenge of cement substitutes. How existing facilities can make new products must be considered.

1.5. Calera process (carbonate mineralization and aqueous precipitation). In the Calera process, CO_2 from flue gases of thermal power plants is mixed with a seawater-brine solution. It bonds with calcium and magnesium in seawater to form carbonates which are precipitated

Designing Green Cement Plants
http://dx.doi.org/10.1016/B978-0-12-803420-0.00048-2

from water and dried using the heat of flue gases. Solidified minerals can be used as cement or as aggregates.

The process mimics the marine cement produced by corals to make shells and reefs.

The process removes calcium and magnesium from seawater and makes its desalination easy. It captures 95% of the SO_2 in flue gases and neutralizes other pollutants like mercury.

CO_2 is a raw material for the process. For every unit of CO_2 that the conventional process emits to atmosphere, Calera absorbs three. Every ton of coal burnt to produce electricity produces 25 tons of CO_2. Calera captures it to make 5 tons of cement or aggregate. It thus saves 7.5 tons of limestone. Attaching the Calera process to a coal-fired thermal power plant is a very promising proposal because its flue gases contain $\sim$15% CO_2. A pilot plant is being installed to prove the process commercially. The Calera process is actually a carbon-capturing and sequestration process. It not only captures carbon dioxide but makes a cement-like product out of it.

1.6. Novacem

Novacem is also known as CO_2-absorbing or carbon-negative cement. It is based on using magnesium silicate (talc) as raw material rather than limestone. This material is abundantly available (10,000 billion tons) and will support large-scale production. Magnesium silicate is converted into magnesium carbonate under elevated temperatures (180 °C) and 150 bar pressure. Carbonate produced is heated at low temperatures of $\sim$700 °C to produce MgO, which is then made into cement. As it hardens within concrete mixes, it actually absorbs and stores atmospheric CO_2. For every ton of Novacem produced, three-fourths of the CO_2 is absorbed. CO_2 is recycled back into the process. Low temperatures permit use of alternative fuels (biomass) which further reduces CO_2 emissions. One ton of Novacem absorbs 100 kg more CO_2 than it emits. Total typical emissions in making Novacem are -50 to $+100$ kg/ton of cement as compared to $\sim$800 kg/ton for OPC. Novacem consumes 60-90% of the energy normally required for OPC. It is claimed that Novacem will be on par with OPC in quality and costs of production. Novacem is naturally white in color. A 25,000-ton capacity pilot plant is being installed.

1.7. Calix

Calix is an Australian company that proposes to make cement using superheated steam which modifies cement particles and makes them chemically more reactive. This process separates CO_2, making it easy to capture it. The process uses a flash calciner. The main product is semidolime (calcined dolomite), which can replace cement.

Horley flash calciner converts lime, dolomite, or magnesite into their respective oxides in a far more efficient process. In the process, fine dolomite particles are dropped down a vertical tube containing superheated steam at 400 °C. They are converted into oxide before they reach bottom in about three seconds. Particles also have a very large specific surface of about 100,000 m^2/kg. Therefore chemical reactions are very fast. Semidolomite binds so well that a Calix product develops the same strength as normal concrete in twenty minutes. Products made from semidolomite retain the CO_2 bound to calcium as $CaCO_3$. CO_2 released from $MgCO_3$ is captured as a pure gas stream which can be sequestered.

A flash calcining plant of Calix became operational in 2007.

1.8. Geopolymer cements (GPC)

There are different types of GPCs.

1. rock based
2. fly ash based

GPCs also use coal mine wastes as base raw materials.

Fly–ash-based GPC uses fly ash from thermal power plants produced at 1200-1400 °C as raw material.

Fly ash	50-80%
Slag	10%
Potassium silicate	10%

Rock-based GPC reduces CO_2 emissions by 80%. It does not require calcination of limestone. Therefore there is no release of CO_2. It requires heating at low temperatures. It requires less capital investment and its properties are comparable to those of OPC.

Appropriate geological resources are available to provide raw materials. GPCs would allow unlimited development of concrete infrastructures and economy. GPCs would mitigate GHG emissions. Some GPCs have been compared with OPC in the following table.

Cement Type	Manuf. Temp. (°C)	Energy	CO_2 Emission
OPC	1400-1500	100	100
Glass	750-1350	64	35
Carbunculus[TM]	750-800	40	20
Carbunculus[TM] nat.	20-80	30	10

Fly–ash–based GPCs reduce CO_2 emissions by 90%.

Geopolymers are synthetic alumina silicate materials with potential for replacing OPC. GPCs have greater thermal and chemical resistance and better mechanical properties. Since no limestone is used, GPC has excellent properties in an acid and salt environment. Seawater can be used to blend GPCs. Geopolymerization starts with an alkalization step involving NaOH and KOH. Fly-ash-based GPC binds with coarse and fine aggregates to form GP Concrete. Geopolymer cement (GPC) will mitigate global warming.

Existing cement plant production facilities can also produce GPC. The same grinding mills and kilns can be used. Kilns run at half the temperature maximum 750 °C Clinkering temperature for GP cement is about half-between 700-750 °C Fuel consumption is reduced by two-thirds. Therefore fuel consumption is reduced by about one third as compared to common cement.

1.9. Aether[TM]

Lafarge has developed a new cement clinker with properties similar to OPC clinker that can be produced in existing cement plants but with much lower (25-30%) CO_2 emissions. Reportedly its composition is

Belite	C_2S 40-75%				
Calcium sulfoaluminate	C4A3$15-35%				
Ferrite: C_2 (A,F)	5-25%				
Minor phases	0.1-10%				
Loss of ignition	Portland raw mix 35% Aether raw mix 29%				
Chemical Composition					
Portland clinker		C_3 S	C_3 A	C_2 S	C_4F
%		65	~6	~15	~12
Aether clinker		$C_4A_3$$	C_2S	C_2 AF	
%		~25	~55	~20	

Aether clinker can be produced in normal cement kilns at sintering temperatures of 1225-1300 °C Performance of concrete made with Aether cement is comparable with that of OPC concrete. Reportedly the energy required for grinding Aether clinker is less than that for OPC.

Thus, on all three counts, that is, CO_2 emission, consumption of thermal energy and consumption of electrical energy Aether is claimed to be superior to normal OPC. Further, Aether clinker can be blended with the usual blending materials like fly ash and slag to make blended cements.

Lafarge have undertaken commercial scale trials in their cement plant in France.

1.10. It is apparent from the details available about the various cement substitutes mentioned above that they are essentially **green cements.** Therefore the cement industry should watch as they become commercially viable products and even contribute to their development. It is in the interests of the cement industry to switch over to new products using their existing production facilities. Only then will green cements be readily available to consumers. To start with, it can be determined if CO_2 from waste gases can be used to make substitutes like the Calera mentioned above as a byproduct.

RECOMMENDED READING & WEBSITES

1 Novacem
   1. Novacem: Carbon negative cement & green cement bond CSI Forum 2010: by Stuart Eavans & N. Vlasopouos
   2. Building a better world with green cement by Romanao

Websites
   1. www.novacem.com/technologies/overview
   2. www.wbcsdcement.org
   3. www.smithsonian.com
   4. www.gizmag.com
2 Calera
   1. Calera - carbon capture and sequestering technology

Websites
   1. www.calera.com
   2. www.scientificamerican.com

3 Calix

   1. Cement Technology Road Map 2009
   2. Looking beyond OPC - Cement Industry News

   Websites

   1. www.calix.au.com
   2. www.iea.org/paper
   3. www.globalcement.com
   4. www.smithsonian.com
   5. www.en.wikipedia.org

4 Geopolymer Cements (GPC)

   Geo polymer Alliance - Prospectus 2009

   Websites

   1. www.geopolymer.org/applications/globalwarming
   2. www.en.wikipedia.org
   3. www.primaryinformatin.com/geopolymercement
   4. www.researchgate.net
   5. www.gizmag.com/green-
   6. www.davidouts.info

5 Aether

   1. Presentation at ECRA
   2. Innovating to reduce Carbon Footprint of Cement Production - Brochure by Lafarge and others
   3. CO_2 New Cements and Binder Technologies- Aeher (BCSAF) Cements presentation by Dr. Walenta & Dr. C. Comparet of Lafarge Research Station, Lyon

   Website

   1. www.aether-cement.eu

SECTION 10

Conclusion

Contents

CHAPTER 1

Conclusion

1.1. Impact of improved technology and efficiency

Though the costs of all raw materials and fuels and services like power and labor have gone up appreciably, the full impact has been mitigated by improved technology and improved operational efficiency.

- fuel efficiency stands at 650-700 kcal/kg
- specific power consumption is between 70 and 80 kwh/ton
- productivity expressed as tons per man hour is around 6.5

1.2. Lower costs of blending materials

Use of blending materials like fly ash and blast furnace slag has not only conserved limestone but has also resulted in reduced costs of production, as shown in Section 8.

CoP	OPC	PPC	BFSC
Rs./ton	1830	1598	1454
$/ton	~33	~29	26.4

1.3. Lower costs of AF and power generated by WHRS

Coal costs on average are Rs. 4500/ton ($82) as received. Waste fuels costs much less, on average Rs. 2000/ton ($36.4)

Similarly grid power costs ~Rs. 5/kwh (~$0.09)

WHR power costs Rs. 2.5/kwh (~$0.045)

These lower costs of AF and WHR power further bring down production costs.

1.3.1. Production costs of OPC, PPC, and BFSC with AF & WHR are shown in Table 8.1.2

CoP with AF & WHR	OPC	PPC	BFSC
Rs./ton	1710	1480	1420
$/ton	~31.1	~26.9	~25.8

It is the trend of these values that is significant, not the figures themselves, which may vary from place to place.

1.4. Lower emission of greenhouse gases and saving of natural resources

The impact of reduction in GHG emissions on account of the higher cement/clinker ratio (using blending materials), alternative fuels and a WHR system has already been described in Chapter 2 of Section 1.

Briefly, CO_2 emissions in kg/kg clinker at the present level of fuel efficiency and limestone to clinker ratio are:

		After AF	After WHR
Emissions due to	Now	in kg/kg clinker	
Calcination	0.51	0.51	0.51
Combustion	0.29	0.27	0.27
Waste heat recovery			−0.03
Total (use of AF ~ 20%)	0.8	0.78	0.75

In terms of cement produced, this is reduced to

Cement	Cement/clinker Ratio	Emission in kg/kg cement		
OPC	1.05	0.76	0.74	0.71
PPC (30% fly ash)	1.54	0.52	0.51	0.49
BFSC (60% slag)	2.86	0.28	0.27	0.26

1.4.1. There is not much scope for further reduction due to savings in fuel consumption. Therefore there is not much scope for reduction on account of WHR.

The real difference in GHG emissions is due to the cement/clinker ratio, that is by blended cements. The cement industry the world over is directing its attention to maximizing blended cements. Already the proportion of OPC in the total production of cement has dropped to about 25%.

1.5. Savings of natural resources like limestone are in direct proportion to the cement/clinker ratio. The percentage of AF used proportionately saves fuel.

1.6. Green cement plants as proposed meet with the above objectives admirably and also reduce costs of production and require less investment as expressed in Rs. ($)/ton of cement.

This fact will spur setting up of green cement plants. Hopefully all new cement plants will be green plants and existing plants will be converted into green plants.

1.6.1. The various benefits have been shown on **a unit basis, per kg or per ton**. Impact on a universal basis can be readily calculated.

Each country can assess its present status of greening and work out the scope and savings possible in future.

Benefits for the world can also be worked out on the basis of world production of different types of cement and the prevailing use of AF and WHR.

1.7. Capital costs

The impact on capital costs of making blended cements using AF and installing WHR is outlined in Section 8 and summarized in Table 8.1.1.

Capital costs expressed in Rs./$ per ton come down from Rs. 4400 (~$80) to Rs. 2580 (~$47) when comparing OPC and BFSC with the use of AF and WHR.

1.8. Future

In the foregoing sections the various ways in which these objectives are achieved and the extent to which they can be achieved have been described.

Even possibilities that appear remote now, like carbon capture and low carbon cements (substitutes) have been touched upon.

1.8.1. Judging by the scope for further reduction in GHG gases by the present methods it is clear that research will have to be directed toward making practices like carbon capture viable for the cement industry.

1.8.2. New cement products like Calera produce cement by using CO_2 from power plants.

Cement companies should actively launch research regarding cement substitutes which promise to be green cements, as suggested in Section 9.

1.8.3. Renewable energy is beginning to make an impact. Several cement plants have invested in setting up wind farms. Solar power at this juncture is much costlier with regard to both investment and production costs. But there are indications that these will come down appreciably in the near future. Lifecycle costs are low for both wind and solar power. Wind turbines and solar panels are guaranteed for 25 years. Thus renewable energy will, sooner rather than later, become an integral part of green cement plants.

1.9. Social impact

An indirect but very beneficial social impact is the mitigation of waste management problems for industry and cities.

Cement substitutes on the horizon are green.

1.10. Beginning to design a green cement plant

Section 7 furnishes pertinent details that will help cement engineers, cement consultants and cement machinery manufacturers take up the design of large-capacity green cement plants.

1.11. Win–win–win situation

Thus it is not just a win–win situation. It is a **win–win–win** situation.

Therefore it should be **"go-go-go"** all the way for green cement in its present form as also in its future form.

1.12. The actual impact varies widely from country to country depending on the present state of greening in that country and the scope for further greening.

Note: conversion rate: 1$ = 55 Rs.

SECTION 11

Sources

Contents

CHAPTER 1

Books & Periodicals

Sr. No.	Title	Author/Publisher/Firm
Books		
1	*Cement Statistics 2011*	Cement Manufacturers' Association
2	*Manual on Waste Heat Recovery*	CII-Sohrabji Green Business Centre
3	*Case Study Manual on Use of Alternate Fuels and Raw Materials in Cement Industry*★	As above
4	*Manual on Best Practices in Indian and International Cement Plants*	As above
5	*Manual on Best Practices in Cement Industry: World Class Energy Efficiency Initiative in Cement Industry*	As above
6	*Low Carbon Road Map for Indian Cement Industry*	As above
7	*Manual on Thermal Energy Efficiency in Cement Industry*	As above
8	*Cement Formulae Handbook Version 2*	As above
9	*Globalization of Indian Cement & Construction Industries*	NCCBM International Seminar 2005
10	*Development of Bulk Packaging Systems using Road Transport*★	NCCBM
11	*Handbook for Designing Cement Plants*	S.P. Deolalkar/BS Publications
12	*Nomograms for Design & Operation of Cement Plants*	S.P. Deolalkar/BS Publications
Periodicals		
1	*Zement Kalk Gips*	Bauverlag
2	*Cement Journal*	CMA
3	*International Cement Review*	Cement
4	*World Cement*	World Cement
5	*Cement Energy & Environment*	CMA

Designing Green Cement Plants
http://dx.doi.org/10.1016/B978-0-12-803420-0.00050-0

CHAPTER 2

Brochures and Catalogs (in print and in CD form)

Sr. No.	Title	Author/Publisher/Firm
	In Print	
1	*Alternative Fuel Solutions*	FLS
2	*Horomill, TSV Separator*	Fives FCB
3	*Competence in Cement*	GEBR/Pfeiffer AG
4	*Hot Disc Technology*	FLS
5	*Cemergy—Waste Heat Recovery Systems*	FLS
6	*Novaflam & Precaflam Burners for Kiln and Calciner*	Fives Pillard
7	*Refurbishment*	FLS
8	*Integrated Architecture Cement Industry Solutions*	Rockwell Automation
9	*Cement Industry: World Class Solutions for Cement Plants*	Siemens
10	*Energy Solutions for a Green World*	Thermax
11	*Waste Heat Recovery Systems for Cement Plants*	Thermax
12	*Electrostatic Precipitators*	Thermax
13	*ACE Jet Bag Filters*	Thermax
14	*Cleaner & Greener World for Tomorrow*	Transparent Technologies Pvt. Ltd
15	*Preheaters & Precalcinators*	PSP Engineering AS
	In CD Form	
	Sets of Brochures for their Cement Plant Products available from	
1		FLS
2		KHD
3		Krupp Polysius
4		Pfeiffer
5		Transparent Energy Systems Pvt. Ltd
6		Clair Engineers Pvt. Ltd
7		Rockwell Automation

Designing Green Cement Plants
http://dx.doi.org/10.1016/B978-0-12-803420-0.00051-2

Brochures & Catalogs Downloaded

Sr. No.	Company	Brochures for
1	ABB Automation	Instrumentation & Process Control
2	Aumund	Materials Handling
3	Crompton Greaves	Motors
4	Cummins	D.G. Sets
5	David Brown	Gearboxes
6	Ecomak	Dust Collecting Equipment
7	Kay Blowers	Roots Blowers
8	McNally Sayaji	Crushers
9	Kirloskar	Motors
10	Pfeiffer	MPS Mills
11	San Engineering	Locomotives
12	Siemens	Process Control
13	Bharat Earth Moving Ltd	Earth Moving Machinery
14	Wartsila	D.G. Sets
15	Thermax	ESPS, Bag Filters, WHR Boilers
16	Claudius Peters	e-mill, clinker cooler, blending silos, packing machines, etc.
17	Polysius	Complete range of cement machinery quality control and process control
18	Hazemag	Full range of crushers & feeders
19	CII	Green Buildings & Homes
20	Rockwool	Insulation
21	Indian Railways	Details of track, wagons, etc.
22	GEPIL	Alternate fuels processing, etc.

We are grateful to the companies above for making available catalogs and brochures.
Note: FLS = FLSmidth Pvt. Ltd.; KHD = KHD Humboldt wedag; Polysius = Polysius Krupp Thysers; Pfeiffer = GEBR Pfeiffer.

CHAPTER 3

Plates

Chapter No.	Plate No.	Title	Source
		Section 3	
1	3.1.1	Schematic diagram of possible CCS systems	Technical Report TR004/2007
		Section 4	
7	4.7.1	Briquetting press	Koppern
		Section 5	
3	5.3.1	Organic Rankine cycle power plant	Ormat International Inc.
4	5.4.1	WHR boiler for clinker cooler	Kawasaki Plant Systems
	5.4.2	WHR boiler for preheater	Thermax
	5.4.3	WHR boiler for clinker cooler	Thermax
		Section 6	
7	6.7.1	Wind power at Tetouan Cement Plant	Lafarge Morocco
8	6.8.1	Concentrated solar power project	Solar Power
		Section 7	
6	7.6.1	Aeration of sectioned silos for storing cements	
	7.6.2	Extraction from different levels in multi-compartment silos	Article by HU Schalkhaar in ZKG
	7.6.3	Arrangement of a cement silo for multiple extraction	Brochure of Claudius Peters
7	7.7.1	Compound primary crusher	Hazemag
	7.7.2	MVR Multi-drive V. R. mill	GEBR Pfeiffer
	7.7.3	V separator	Humboldt Wedag
	7.7.4	Roller press	Humboldt Wedag

Continued

Designing Green Cement Plants
http://dx.doi.org/10.1016/B978-0-12-803420-0.00052-4

Chapter No.	Plate No.	Title	Source
	7.7.5	Tangential tire suspension system	FLS
	7.7.6	Friction drive-through supporting rollers	FLS
	7.7.7	Self-aligning roller support	FLS
	7.7.8	Duoflex multi-channel burner	FLS
	7.7.9	Hot Disc for feeding whole tires	FLS
	7.7.10	Gunnax system for whole tires	FLS
	7.7.11	Gunnax firing waste canister	FLS
	7.7.12	Low NO_x calciners	FLS
	7.7.13	Concept of walking floor for coolers	Claudius Peters AG
	7.7.14	Polytrack clinker cooler	Polysius Knepp Thysson

Note: FLS—FLSmidth Pvt. Ltd.

CHAPTER 4

Figures

Chapter No.	Fig. No.	Title	Source
		Section 3	
1	3.1.1	Flow chart for oxyfuel firing process	Carbon Capture Technologies
	3.1.2	Flow chart for amine scrubbing process	PCA R&D Serial 3022
	3.1.3	Flow chart for carbonation calcination loop	Greenhouse Gases Science & Technology
		Section 4	
5	4.5.1	Flow chart for acceptance and rejection of wastes proposed for use	CII—Case study Manual for AF&R
	4.5.2	Steps to prepare wastes for use as AF by ISP	GEPIL—Waste preparation facility
	4.5.3	Steps to prepare AF from wastes in greater detail	As above
6	4.6.4	Flow chart for preparing waste sludge for use as AF	ANZBP Hobart Road Show 2011
8	4.8.1	Savings in GHG emissions	AF in cement manuf., Technical & Environmental Review a Cembureau Presentation
		Section 5	
6	5.6.11	Flow chart for WHR system showing two turbines on same shaft	NCCBM Presentation see References Section 5

Continued

Designing Green Cement Plants
http://dx.doi.org/10.1016/B978-0-12-803420-0.00053-6

Chapter No.	Fig. No.	Title	Source
	5.6.12	Flow chart for common turbine and separate WHRBs for Preheater & Cooler	Kawasaki Plant Systems See references in Section 5
	Section 6		
7	6.7.1	Relation between wind speed and power generated	Brochure of Suzlon Energy Ltd
	6.7.2	Simple scheme of grid-connected wind power plant for a cement plant	Wind generation application for Cement Industry
8	6.8.1	Circuit diagram for a PV solar power plant	Report IISC-DCCC 11 RE 1
	6.8.2	Common types of solar power systems	Science & Technology of Photovoltaics by P. Jayarama Reddy
	Section 7		
7	7.7.2	Control of NO_x by multiple feeds and injections of ammonia	VDZ Activity Report 2003-2005

CHAPTER 5

Tables and Annexures

Chapter No.	Table/ Annexure No.	Title	Source
		Section 4	
1	4.1.1	Desirable properties of AFs	★ not certain
2	4.2.1	Typical calorific values of different fuels and AFs	CII—Case Study Manual for AF&R
	4.2.2	Properties of some common wastes	As above
	4.2.3	Typical composition of municipal solid waste (MSW)	Article by Axel Seemann Centre for sustainable development in Cement Industry
	4.2.3a		US Environmental Protection energy
	4.2.3b	Typical composition of MSW in India	National Solid Waste Association of India
	4.2.4	Wastes not suitable as AFs	CII—Case Study Manual for AF&R
8	4.8.1	Brief data on some AFR installations in India	CII—Case Study Manual for AF&R
9	4.9.1	Alternative raw materials	CII—Case Study Manual for AF&R
		Section 5	
3	5.3.1	Salient features of various WHR systems	NCCBM Presentation See references in Section 5

Continued

Designing Green Cement Plants
http://dx.doi.org/10.1016/B978-0-12-803420-0.00054-8

Chapter No.	Table/ Annexure No.	Title	Source
		Section 6	
2	Annexure 1	Norms & guidelines for green buildings	CII -Publications IGBC Green Buildings Rating System Pilot Version 1.0IGBC Green Homes Rating System Version 1.0
3	Annexure 1	Conditions for sanctioning a cement plant project	Extracts from a letter of Consent from MoEF
7	6.7.1	Performance data on wind power	Annex III, Recent Renewable Energy Costs & Performance
8	6.8.1	Performance data on solar power	University of Cambridge As above
		Section 7	
7	7.7.1	Quarry equipment for a 1000-tpd cement plant	B.E.M.L. Brochures
	7.7.2	Two, three support kilns	FLS Brochure on Dry Process Kilns
	7.7.3	Kiln capacity & number of preheater streams	As above

INDEX

Note: Page numbers followed by *t* indicate tables.

Printed in the United States
By Bookmasters